AAPS Advances in the Pharmaceutical Sciences Series

Volume 26

The AAPS Advances in the Pharmaceutical Sciences Series, published in partnership with the American Association of Pharmaceutical Scientists, is designed to deliver volumes authored by opinion leaders and authorities from around the globe, addressing innovations in drug research and development, and best practice for scientists and industry professionals in the pharma and biotech industries.

More information about this series at http://www.springer.com/series/8825

Mario L. Rocci Jr. · Stephen Lowes
Editors

Regulated Bioanalysis: Fundamentals and Practice

Springer

Editors
Mario L. Rocci Jr.
Laboratory Services
ICON, Plc
Whitesboro, NY
USA

Stephen Lowes
Bioanalytical and ADME Labs
Q Squared Solutions
Ithaca, NY
USA

ISSN 2210-7371 ISSN 2210-738X (electronic)
AAPS Advances in the Pharmaceutical Sciences Series
ISBN 978-3-319-54800-5 ISBN 978-3-319-54802-9 (eBook)
DOI 10.1007/978-3-319-54802-9

Library of Congress Control Number: 2017933942

Printed on acid-free paper

This Springer imprint is published by Springer Nature
The registered company is Springer International Publishing AG
The registered company address is: Gewerbestrasse 11, 6330 Cham, Switzerland

To my beloved wife and son, Donna and Mark Rocci for all the love, encouragement and support they've provided throughout my professional career.

—Mario L. Rocci Jr.

To Kathy, Thea and Elise Lowes—my loving family and all I hold dear.

—Stephen Lowes

Preface

Bioanalysis, simply described as the quantitative measurement of the concentration of a drug, its metabolite(s), or an endogenous compound present in a biological matrix, is essential to the discovery, development, and regulatory approval of new medicines. It is also key to modern pharmacokinetics (i.e., the study of what the body does to a drug) and, increasingly, to pharmacodynamics (i.e., the study of what a drug does to the body). Understanding these biological processes is critical to the practice of determining drug safety profiles and efficacy potentials as well as to establishing recommended dosage for new therapeutics that treat disease. Generating this important data to get safe and effective medicine to patients is the work of contemporary bioanalysts. It is the ingenuity of the bioanalysis team that enables development of robust assays to meet increased sensitivity and selectivity requirements across a wide range of chemical entities, from small molecule drugs to protein- or DNA-based therapeutics.

Once considered a branch of analytical chemistry, bioanalysis now spans biology and chemistry. It shows no sign of limitations in furthering drug development in, either the needs it can address or its potential to advance on those requirements. The advances in technology and strategies implemented in today's bioanalytical laboratory are in accordance with the challenges and opportunities presented by new drug therapies. Despite the developments in bioanalytical technology available to today's bioanalyst, an understanding of the fundamentals of this quantitative analytical science remains incredibly important. Whether setting up a new lab, reorganizing an existing lab, considering an investment in new technology or software, writing SOPs or defending data to an auditor, a firm grasp of the fundamentals will serve the bioanalyst well. While there is currently a wealth of information available on advanced bioanalytical approaches, the volume of material can be overwhelming to those new to or considering a career in bioanalysis. Add to this the evolving regulations guiding practice, and the need for a fundamental reference on modern bioanalysis arises, inclusive of the science, technology, and regulations. It is this need that prompted us to approach recognized bioanalysis experts and opinion leaders to write this book. With it, we aim to provide a comprehensive text that puts fundamental bioanalytical science in context with current pharmaceutical bioanalysis practice, its challenges, and ongoing developments. It expands on existing texts on the subject by covering regulated bioanalysis of both small and large molecule therapeutics. We hope the content will be useful to a wide spectrum of readers: from those new to bioanalysis, to those developing their experience in the laboratory or working in one of the many critical supporting roles, to seasoned practitioners looking for a solid source of information on this exciting and important discipline.

Chapter 1 provides an introduction to the book. Chapter 2 is a primer on regulations affecting bioanalysis, describing what global health authorities require and expect of organizations generating bioanalytical data for new drug applications. An important foundation for the discipline, it also points to the challenge of drafting prescriptive and comprehensive guidance in a variable and rapidly evolving scientific environment. Chapter 3 summarizes the logistics and practice of establishing and running a bioanalytical laboratory. Chapter 4 discusses the documentation needed for such an endeavor. Chapters 5–7 focus heavily on small molecule bioanalysis and liquid chromatography-mass spectrometry (LC-MS) applications, including how LC-MS is now finding niche applications beyond small molecules, including for large molecule and biomarker bioanalysis. Ligand-binding assays (LBA) are noted as the go-to analytical approaches for peptide, protein, and other biomolecule work since the introduction of immunoanalytical techniques. With the surge in protein therapeutics entering pharmaceutical and biotechnology pipelines, we expect this trend to continue. Chapters 8 and 9, therefore, focus on LBA techniques, and also discusses where and how the regulations apply in light of the nuances of these approaches.

This book was made possible by the dedication of the chapter authors. We are humbled by their generosity, both in time and effort, in sharing their valuable knowledge with the bioanalytical community. It is with deepest gratitude that we thank them for contributing excellent first drafts and for sticking with us through rounds of review. These thanks extend to their families, whom we know routinely accept this extra, uncompensated effort beyond the "day-job."

We also would like to thank Kathryn Henion, Ph.D., for her valuable proofreading edits. A third set of eyes and a masterful red-pen helped tremendously in pulling this book together. Finally, we would like to thank Springer Publishing for helping us through the process of bringing this book to fruition and into the hands of readers who, we hope, will benefit from the content. Thank you all.

Whitesboro, NY, USA Mario L. Rocci, Jr.
Ithaca, NY, USA Stephen Lowes

Contents

Editors and Contributors

About the Editors

Mario L. Rocci Jr. is Senior Vice President at ICON Laboratories and was formerly the CEO of Prevalere Life Sciences, Inc. He has held appointments in the Department of Medicine at Jefferson Medical College, and Pharmaceutics and Pharmacy Practice at the Philadelphia College of Pharmacy and Science.

Dr. Rocci has a B.S. in Pharmacy and a Ph.D. in Pharmaceutics. He is a Fellow of the American College of Clinical Pharmacology (ACCP) and the International Pharmaceutical Federation (FIP). Dr. Rocci is a past President of the American Society for Clinical Pharmacology and Therapeutics (ASCPT) and the American Association of Pharmaceutical Scientists (AAPS), where he received the AAPS Distinguished Service Award. He is a past Chairman of the Board of the Product Quality Research Institute (PQRI) and a past Board Member of the International Pharmaceutical Federation (FIP). Dr. Rocci also serves as Chair of the Dean's Advisory Council in the School of Pharmacy and Pharmaceutical Sciences at the University at Buffalo, where he received the Willis G. Gregory Memorial Award, and is a member of the External Advisory Board of the University of Kentucky's, School of Pharmacy.In addition to his accomplishments in the industry, Dr. Rocci has authored over 150 publications.

Stephen Lowes is a Senior Director at Q^2 Solutions. Dr. Lowes has over 25 years of experience in bioanalysis with a focus on LC/MS and regulatory requirements. He currently leads Q^2 Solutions' regulated LC/MS bioanalytical CRO laboratory, supporting drug development initiatives of global pharmaceutical and biotechnology sponsors.

Dr. Lowes received his bachelor's degree in Analytical Chemistry at Sheffield Hallam University (UK) and Ph.D. in Biochemistry at The Open University, Milton Keynes, (UK). He is a steering committee member and past-chair of the AAPS bioanalytical focus group, is a founding member of the Global Bioanalysis Consortium (GBC) and is an active contributor to the Global CRO Council (GCC). Dr. Lowes is a regular presenter, trainer and author of publications on regulated bioanalysis with emphasis on LC/MS approaches.

Contributors

Mark Arnold Burlington, NJ, USA

Surendra K. Bansal Roche Pharma Research and Early Development, Pharmaceutical Sciences, Roche Innovation Center New York, New York, NY, USA

Patrick Bennett PPD Laboratories, Richmond, VA, USA

Ronald R. Bowsher B2S Life Sciences, Franklin, IN, USA

Sheila Breidinger Merck Research Labs—PPDM Clinical Bioanalytics, West Point, PA, USA

Stephanie S. Cape Covance Laboratories, Inc., Madison, USA

Aleks Davis B2S Life Sciences, Franklin, IN, USA

Binodh DeSilva Analytical and Bioanalytical Operations, Bristol-Myers Squibb, Princeton, NJ, USA

Douglas M. Fast Covance Laboratories, Inc., De Forest, USA

Van Fitzgerald Radix BioSolutions, Georgetown, TX, USA

Eric Fluhler Washingtonville, NY, USA

Boris Gorovits Biomedicine Design, Pfizer Inc., Andover, MA, USA

Jack Henion R&D, Q Squared Solutions, Ithaca, NY, USA

Charles Hottenstein Bioanalysis, Immunogenicity and Biomarkers, GlaxoSmithKline, King of Prussia, PA, USA

Mohammad R. Koupaei-Abyazani Covance Laboratories, Inc., Madison, USA

Richard M. LeLacheur Agilux Laboratories, Inc., Worcester, MA, USA

Stephen Lowes Bioanalytical and ADME Labs, Q Squared Solutions, Ithaca, NY, USA

Min Meng Covance Laboratories, Inc., Salt Lake City, USA

Johanna Mora Analytical and Bioanalytical Operations, Bristol-Myers Squibb, Princeton, NJ, USA

William Nowatzke Radix BioSolutions, Georgetown, TX, USA

Mario L. Rocci Jr. Laboratory Services, ICON, Plc, Whitesboro, NY, USA

Gary Schultz Munson Technology LLC, Ithaca, NY, USA

Franklin Spriggs Ligand Binding Assays, AIT Bioscience, Indianapolis, IN, USA

Faye Vazvaei Roche Pharma Research and Early Development, Pharmaceutical Sciences, Roche Innovation Center New York, New York, NY, USA

Eric Woolf Merck Research Labs—PPDM Clinical Bioanalytics, West Point, PA, USA

Chapter 1
Introduction

Stephen Lowes and Mario L. Rocci Jr.

Abstract Bioanalysis spans nearly a century of practice enabling the widespread use of pharmacokinetics and pharmacodynamics to improve the rational use of existing drugs and the development of new drugs. While few universities and colleges offer formal programs in bioanalysis, there are several scientific degree programs that are good entry points into the field. The successful bioanalytical laboratory requires a high degree of scientific talent in addition to professionals capable of tackling operational and sample logistics, project and data management, quality control and quality assurance, metrology, instrument maintenance, information technology, and technical writing. These positions and others afford attractive career opportunities for those contemplating the field. Technology is developing at an amazing rate in the bioanalysis field and it is an exciting time to be engaged in this scientific discipline. Given this rapid pace of change, keeping current with the scientific literature, as well as active participation in professional organizations, focus groups, and scientific conferences and meetings germane to the field is crucial.

Keywords Bioanalysis • Bioanalytical career opportunities • Bioanalytical laboratory personnel • Bioanalytical information sources

---

S. Lowes (✉)
Bioanalytical and ADME Labs, Q Squared Solutions,
19 Brown Road, Ithaca, NY 14850, USA
e-mail: stephen.lowes@q2labsolutions.com

M.L. Rocci Jr.
Laboratory Services, ICON, Plc, 8282 Halsey Road, Whitesboro, NY 13492, USA
e-mail: Mario.rocci@iconplc.com

M.L. Rocci Jr. and S. Lowes (eds.), *Regulated Bioanalysis: Fundamentals and Practice*, AAPS Advances in the Pharmaceutical Sciences Series 26,
DOI 10.1007/978-3-319-54802-9_1

1.1 History, Background, and Status of Regulated Bioanalysis

Bioanalysis spans nearly a century of practice applied in pursuit of new medicines for human and animal use. The quantitative measurement of drugs and their metabolites in biological fluids, in addition to endogenous markers of drug effects, enabled the fundamentals of pharmacokinetics (PK) and later pharmacodynamics (PD)—two disciplines critical to the discovery and development of new drugs [1, 2].

Bioanalytical methods are developed and validated under very challenging conditions. Quantifying a drug, its metabolite(s) or endogenous analytes of interest is complicated by the biological sample itself, which often contains very high levels of structurally similar and naturally occurring endogenous compounds. Furthermore, these measurements must be made at very low levels and with a high degree of accuracy and precision. As drug development has evolved over the last few decades, the types of molecules undergoing development have become more diverse. Most notable has been the evolution of biotechnology which has changed the therapeutic landscape to include the development and use of protein therapeutics to treat diseases of all types. This has required bioanalytical science to broaden its reach to include the analysis of macromolecules in addition to small molecule drugs. As we will cover in several chapters in this book, different technology and approaches are necessary for analyzing low-molecular-weight, small molecule compounds compared to much larger, protein-based entities.

Today, bioanalytical assays are typically divided into two categories: chromatographic assays and ligand-binding assays (LBA). These categories can be further divided into the techniques used to perform them. While liquid chromatography–mass spectrometry (LC-MS) based techniques dominate chromatographic approaches, immunoanalytical techniques have enjoyed a tremendous renaissance in a wide range of LBA approaches. More recently, both LBA and LC-MS are being employed together in hybrid assays as a selective, sensitive, and adaptable approach for the bioanalysis of the latest (and, arguably, some of the more complex) drugs undergoing development. Commensurate with these technological advances are the evolving and increasingly more sophisticated methods for preparing samples prior to analysis. Pre-analytical activities and sample preparation (e.g., sample collection, extraction, and dilution) are integral to analyte quantitation and must be appropriate for the assay purpose and the detection technology being employed. Whatever the method or technology used, the goal remains the same: to obtain reliable measurements of the physiological concentrations of the analyte(s) of interest at the time of sampling.

Alongside its scientific developments and challenges, bioanalysis has become an area of drug development heavily scrutinized by regulators. This was not always the case. Prior to 1990, there was no specific regulatory guidance to ensure consistency within and between laboratories conducting bioanalytical work. While Good Laboratory Practices (GLP) provided an operational foundation for the discipline, the performance characteristics of a given assay were left to the bioanalyst to define, demonstrate, and defend. Consequently, bioanalytical experiments and strategy varied among laboratories, resulting in a lack of consistency in drug approval submissions. The situation led several experts in bioanalysis, representatives of the US Food and Drug Administration (US FDA, FDA), and the American Association of Pharmaceutical Scientists (AAPS) to organize a workshop in 1990. Held in the Crystal City area of Arlington, Virginia, the meeting was subsequently referred to as the Crystal City workshop on bioanalysis, the first in a series of meetings that have followed. From the first workshop, a conference report was published in 1992 [3] that outlined a consensus view on what experiments and acceptance criteria should be employed when validating a bioanalytical method. While initially this paper did not meet everyone's expectations for bioanalytical method validation (BMV), it did gradually gain acceptance as a standard by both bioanalytical laboratories and regulators (particularly in the USA).

Ten years after the first Crystal City meeting, the bioanalytical community met again for the appropriately named workshop, "Bioanalytical Methods Validation—A Revisit with a Decade of Progress." The resulting workshop conference report [4] was essentially replicated in 2001 as a regulatory guidance issued by the US FDA [5]. Since that time, there has been a continual evolution in both laboratory practice and the expectations from regulators. Today, at least five global health authorities have released their own BMV guidances. While there is a general level of consistency among guidances, further opportunities for harmonization exist within and are desired by the bioanalytical community. Pursuit of such consistency is likely to continue into the foreseeable future, with the ideal objective being a single, globally recognized BMV guidance.

Because less-than-reliable methods can have significant implications with regard to drug safety and therapeutic efficacy, fully validated bioanalytical assays are a critical part of meeting the needs of pivotal PK studies such as bioequivalence (BE) and bioavailability (BA) trials. The critical nature of these assessments has led bioanalysts to adopt comprehensive method validation requirements for all bioanalytical methods used in the drug development setting, without regard to the purpose(s) for which the method is being used. More recently, scientists have instead directed efforts toward considering what level of method validation is suitable and sufficient for an assay's intended use(s). This is termed "fit-for-purpose" or "tiered" approaches.

It is worth noting that bioanalytical regulatory guidance was originally drafted around chromatographic assays, primarily LC–MS. While a section of the 2001 FDA BMV guidance [5] dealt with LBA, the content was insufficiently developed for practical use. Consequently, separate consortia and opinion leaders published highly regarded white papers on the development and validation of bioanalytical LBAs that in turn influenced common practice more so than the official regulatory guidance [6–8]. The Japanese Ministry of Health, Labour, and Welfare (MHLW) is the only global health authority to date to issue a separate BMV guidance for large molecule LBA assays [9, 10], while others follow the FDA's approach of a using separate sections of a common document. Despite continued development and new releases of global health authority guidances, LBAs are still arguably underrepresented in the regulations. With an increasing percentage of biologic drug therapies filling pharmaceutical company pipelines, the call for agreement on what constitutes regulated LBA bioanalytical practice has reached a new urgency. The challenge here of course is that the accelerating pace of technology and science development makes comprehensive and prescriptive regulatory guidance a difficult expectation.

1.2 Roles, Responsibilities, and Career Opportunities in Regulated Bioanalysis

It is without doubt an exciting time to be in the field of bioanalysis. Technology is developing at an amazing rate for both chromatographic and LBA, as well as for hybrid methods that encompass both. Therapeutic drug classes are evolving similarly, creating ever-increasing bioanalytical challenges. Most recently, regulated bioanalysis has been applied to endogenous biomarker assays, further establishing a need for new strategies and approaches.

The successful bioanalytical laboratory requires talent with more than the fundamental science and technology skills. To be sure, an optimally functioning laboratory relies on multidisciplinary teams working in concert toward a common goal: establish robust methods that produce high-quality bioanalytical results. So what are the varied skills and roles that enable a regulated bioanalytical laboratory to thrive? First, is a need for creative bioanalytical scientists with strong scientific knowledge who can develop, validate, implement, and troubleshoot bioanalytical methods. Additional essential support functions are those tackling operational and sample logistics, project and data management, quality control and quality assurance, metrology, instrument maintenance, information technology (IT), and technical writing. In a successful bioanalytical laboratory, all of these professionals must work effectively, efficiently, and collaboratively, within a robust Quality Management System (QMS).

One concern within the modern bioanalysis community is insufficient "new-blood" joining the discipline to meet demand. The reasons for this may be speculated, but they likely relate to regulated bioanalysis being an analytical science niche, with unique requirements that do not fit well within conventional college course curricula or translate directly from other analytical sciences. Even at the graduate level, few research programs directly train and link the intricacies of quantitative bioanalysis with the regulatory requirements that guide this work. In response to this discrepancy, we present a discussion of which college programs of study might fit well with a future career in bioanalytical science. We then describe a few key professional roles within a regulated bioanalytical work environment, each of which provides a stimulating and rewarding career path for those interested in bioanalytical science and technology. These key roles are not inclusive of every aspect of a modern bioanalytical laboratory and other roles offering rewarding career opportunities exist but are beyond the scope of this chapter.

1.2.1 College/University Programs of Study

While much of the education and training necessary for work in a bioanalytical laboratory comes from on-the-job training and experience, there are many programs of study within colleges and universities that can provide a solid foundation for entry into the discipline.

Programs in chemistry, and more specifically analytical chemistry, as well as the pharmaceutical sciences, will provide a solid knowledge base for working in the area of small molecule bioanalysis, which primarily involves quantitative PK assessments. The programs of study affording firm starting points for large molecule bioanalysis are broader given the many techniques and multiple assessments performed in this area. In addition to quantitative PK assessments, immunogenicity and neutralizing antibody assessments are required during the development of protein therapeutics (all of these assessments will be discussed in chapters that follow). As a result, programs in biology, molecular and cell biology, biochemistry, and pharmaceutical sciences are well suited for those interested in large molecule bioanalysis.

The chapters that follow provide a good introduction to bioanalytical sciences including current practice and ongoing advances in technology. We encourage those interested in careers in the field to consider where their education, training, and experience align with this information and where gaps in skill or knowledge exist that can be remediated through additional education or training.

1.2.2 Bioanalytical Laboratory Scientists

Those first being introduced to modern bioanalysis may be forgiven for thinking it is a discipline with rules and regulations that override the science. While bioanalysis does involve regulation, health authority guidance and comprehensive documentation, the foundation of bioanalytical methods relies on sound analytical science and scientists. Bioanalytical scientists are the lifeblood of the laboratory and no laboratory can be successful without progressive, creative, and highly productive scientists. There are a number of scientific roles required within the laboratory. Method development scientists are needed to develop methods from scratch or adapt existing methods, making them more suitable for their intended purpose. In many cases, method development scientists are developing methods for molecules that have never been measured in a particular biological matrix before. This process is extremely challenging, requiring a high level of creativity and scientific knowledge. Because each bioanalytical method is essentially a customized solution to a given analytical need, bioanalysis presents both scientific and operational challenges on an ongoing basis. While common techniques, equipment, and instrumentation may be employed, each bioanalytical method is unique in the combination of steps developed to prepare and analyze samples. In many cases, early feasibility and development experiments need to be conducted to obtain a scientific understanding of the task. These should establish which specific bioanalytical skills will be required, including technique and instrumentation.

Each project starts with understanding the questions that need answering, including but not limited to

Why a particular analyte?

What is already known about the analyte?

What sensitivity of analysis is required?

How stable is the analyte in a collected sample?

What will be the biological matrix?

How will the data be used to make decisions on the development of a drug?

These questions must be answered, as they will influence the bioanalytical approach needed and lead to determining appropriate people, equipment, and time resources for the project.

Once a method is validated, to ensure it is fit for its intended purpose and that it will meet regulatory guidelines, bioanalytical scientists are then responsible for analyzing the thousands of samples generated from drug development efforts. This analysis should be highly efficient and of very high quality, as the data from many of the studies will be submitted to regulators for review as part of a marketing approval application [in the United States a New Drug Application (NDA) or, for a generic drug, an Abbreviated New Drug Application (ANDA)]. For particularly high-volume analyses, automated approaches may be validated and applied to good effect. Introducing automation in a bioanalytical method also requires additional skills and some unique operational specifics that will be covered in several chapters of this book.

1.2.3 Project Leaders and Project Managers

Bioanalytical project management is a multifaceted process. Once the scientific challenge is evaluated, the operational aspects of providing sufficient and appropriate resources become paramount. Personnel, equipment, and allocated time all influence the chances for project success. Experienced bioanalytical project leaders understand these needs and can plan accordingly, factoring in frequently required contingencies. Project leaders often graduate from the laboratory where they have experienced the challenges of developing, validating, and applying bioanalytical methods to study sample analysis. They are typically well organized and have strong documentation and communication skills. Project leaders also need to have an ability to see the big picture and coordinate the efforts of multiple people and teams. Often project leaders juggle multiple projects simultaneously. This role usually involves direct communication with key stakeholders, the scientific teams, support functions, Quality Assurance (QA), and laboratory management.

Project leaders are the key to coordinating remediation efforts when issues arise. From scientific investigations of batch failure to addressing sponsor and QA observations and findings, the project leader is often the single point of control for a regulated bioanalysis project. Likewise, at times of QA audit or regulatory authority inspection, an effective project leader will greatly assist the process. Good project leaders are in high demand, as they are a key influence on the success and reputation of a bioanalytical laboratory.

Where multiple drug development protocols or programs are active simultaneously, it may be necessary to engage a project manager to handle the nonscientific aspects of a series of ongoing projects/programs. This helps ensure the logistical aspects of projects are reasonable, communication channels with all stakeholders are active, effective, and documented, and that satisfactory completion of the work is done according to agreed-upon timelines. In these instances the project leader(s) and project manager work in concert. The project leader focuses on the science, timing, and compliance aspects of the studies (or a subset of them if multiple project leaders are engaged). The project manager works to ensure that all logistics are in place to get the projects/programs completed on time and that communication of project status and issues that arise is quick and effective. Project managers can come from the laboratory or be professionals trained (and often times certified) for this role.

1.2.4 Laboratory Managers

The challenges presented to project leaders and project managers extend to the laboratory manager or management team. Laboratory managers also provide critical support to bioanalysis projects, though this is typically done by ensuring sufficient availability and appropriate allocation of resources. They are also responsible for maintaining an environment conducive to quality and regulatory compliance. When quality issues arise, it is the responsibility of laboratory managers, project managers, and project leaders to ensure that the issues are remediated, as appropriate, through Corrective and Preventative Action (CAPA) plans. Aside from these activities, managers are often responsible for meeting or exceeding their financial budgets or targets, as well as optimal laboratory performance as measured by a variety of metrics (e.g., as discussed later in Chap. 3).

For small, in-house bioanalytical operations, there may be only one laboratory manager coordinating people, equipment, and time allocation directly with the project leaders, project manager(s), or analysts. At the opposite end of the spectrum, within a large Contract Research Organization (CRO), separate departments may exist for each specialized function, each led by a member of the management team. A potential danger here is the development of functional silos each focused primarily on its own interests. Companies cannot afford for this to happen because flexibility and adaptability are required to meet the requirements of varied internal and external stakeholders. It is imperative that laboratory management recognizes and remains sensitive to supporting interdependence among key functions such as sample management, bench scientists, metrology, IT, resource scheduling, and technical writing. This focus fosters collaboration and open communication across the different functions.

1.2.5 Quality Assurance

The QA organization in a regulated bioanalytical laboratory ensures compliance with applicable regulations and guidelines. Despite its importance, QA rarely gets addressed in scientific texts, but valuable and important careers can develop within this discipline.

QA must be kept independent from the operational and support groups within the laboratory. The GLP regulations require a reporting structure within QA that is separate from any personnel or functions that generate data. The role of QA is to ensure performed work complies with regulations where necessary, and to audit the work for compliance with the laboratory's standard operating procedures (SOPs) or any other documented plans, methods, and processes. QA also functions proactively and collaboratively with operational groups to provide input on how those teams might best resolve quality issues. It is therefore important that QA personnel have a basic understanding of the science as well as a deep understanding of the regulations that apply. In this way, remediation of quality issues can occur expeditiously. QA typically leads any external quality visits, audits, and inspections, including those conducted by regulatory authorities. For project-based audits or inspections, all phases of a bioanalytical project may come up for review. This includes review of organizational charts, the master schedule for GLP studies, SOPs, training records, facility records, instrument maintenance and calibration records, project data, archive facilities, and IT infrastructure. In contrast, system-based audits may take a more detailed look into one or more operational systems critical to the laboratory operation.

During inspections, a laboratory tour allows the auditor to correlate practice with written processes and procedures. It is common for such lab tours to be hosted by QA and follow an imaginary sample from receipt through analysis and data reporting. As such, the QA representative must have a good understanding of bioanalytical science, the BMV guidances and associated interpretations of the regulations to manage the process.

1.2.6 Other Roles

The collective effort necessary for a well-functioning bioanalytical laboratory means that other, nonscientific support groups and individuals may benefit from an awareness and understanding of the topics covered in this book. To varying degrees, a working knowledge of regulated bioanalysis helps employees communicate and collaborate with one another. This knowledge builds team respect and appreciation among coworkers. It can also add to job satisfaction. While we do not suggest that every position requires a full understanding of the topics covered in this book, we hope readers from a broad spectrum of job functions across the pharmaceutical industry can find some of the content practical, valuable, and actionable.

Regulated bioanalytical laboratories usually employ some form of quality control (QC) in addition to and separate from QA. QC may vary from within-team peer review of documentation to a fully staffed QC department. In either case, effective QC can have a positive impact on data quality and compliance. It is always better to catch errors early versus later in the process. If the QC function is organized as a dedicated role, then knowledge and understanding of the science and regulations as described in this book will be an asset to job performance.

Similarly, those tasked with the technical writing of reports, research papers, and marketing material should also have a working knowledge of bioanalysis and regulatory expectation so that they may be better able to raise good questions and join dialogue on the topic area. Experienced and knowledgeable technical writers who can spot deficiencies and provide constructive input improve the final product and quickly become a valuable asset to the team.

Business Development (i.e., the group that sells bioanalytical services to new and existing clients) is a key area within CROs. Considering the direct interaction these professionals have with prospective and existing sponsors, it is a tremendous advantage if they can converse on a technical level. Doing so builds credibility and trust and helps ensure that accurate information gets to the bioanalyst.

Health and safety staff, facilities management personnel, and any persons involved in equipment metrology may be called upon during audits and inspections regarding the regulated aspects of their role in the laboratory. A firm understanding of the applicable regulatory requirements for their roles is critical and, in some cases, beyond the scope of this book.

Finally, IT departments face specific regulations (e.g., US 21 CFR Part 11 on electronic records and electronic signatures) requiring compliance in a regulated bioanalytical laboratory. IT staff members are often called upon during audits and inspections to defend practices that ensure the security and authenticity of information and data associated with a regulated bioanalysis project. We therefore advise organizations to train IT personnel in the regulatory requirements as outlined in this book.

1.2.7 Sponsors, Clients, and Data Customers—A CRO Perspective

Those in need of bioanalytical data benefit from understanding the basics of the work they commission a laboratory to perform. Even with the extensive regulations and health authority guidances, the specific nuances of any project call for agreement and communication between the sponsor of the work and the bioanalytical team performing it. While contracting bioanalytical work to a CRO may seem a straightforward process, experience pays dividends both on behalf of the sponsor and the CRO. As will be discussed in the following chapters, many factors contribute to a successful bioanalytical project. The more specific and relevant information a sponsor can provide during the initial contracting stages, the better chance there is for an optimum outcome. In the circumstances of a client outsourcing bioanalytical work to a CRO, experience and good communication skills on both sides of the partnership is especially important. A client who understands the key aspects of establishing and implementing bioanalytical methods, and who can contribute such experience to the project, is a significant advantage. As an independent entity, the CRO often relies on the client to provide key background information. When issues occur at the chemistry and biochemistry level during any phase of the project, the value of a strong client/project manager relationship becomes evident.

An often overlooked stakeholder in a bioanalytical project is the subsequent user of the final data. Typically this is the pharmacokineticist tasked with interpreting the bioanalytical data. Increasingly, it is important to interface with the PK scientist during the planning stages of the bioanalytical work. With bioanalysis now needing to address new compound classes, technology developments, and drug development strategies, it is important that the end-user of the data helps define, at the beginning of a bioanalytical program, what is needed from the method so that the data obtained is most useful for decision-making purposes.

1.3 Beyond This Book

1.3.1 Resources

This book is intended to offer an important foundation in bioanalytical practice. Once the reader acquires a basic understanding of the science, operational, and regulatory requirements on the subject, we encourage them to seek additional information continually in order to stay current.

There are several existing bioanalysis texts that supplement this book and provide additional detail of technology and technique. An online search will lead to titles on principles of bioanalysis, or that focus on technology-specific applications. We view such books as complementary to what we have presented here, often providing additional compound-class case studies or in-depth assessment of instrument type. Several existing texts are focused on small molecule bioanalysis with LC–MS or other instrumental approaches. Others do not place emphasis on regulated bioanalysis and the associated disciplines. Currently, there is a lack of directly relevant books for those looking to advance their understanding of LBA in regulated bioanalysis.

There are several traditional journals that emphasize bioanalysis including *Analytical and Bioanalytical Chemistry* (Springer) and *Bioanalysis* (Future Science) that are tremendous resources for case studies and current developments within the field. Many other well-known analytical chemistry periodicals also publish papers on bioanalytical topics. There are also emerging open-access, peer-reviewed, online journals either focused in this area or with bioanalytical relevance.

We advise the reader to stay aware of ongoing developments through traditional peer-reviewed publications but also to conduct wider online searches of other information sources occasionally. Combined with participation in bioanalytical community discussion forums (see below), there is a wealth of constantly-emerging and valuable information available to the modern bioanalyst with access to the Internet.

The development of regulatory guidance is an ongoing process around the globe. Some of the most pivotal and interesting changes in bioanalytical practice have evolved from the interpretation of the BMV standards. Prior to a guidance becoming effective, a draft form is typically released for public comment. In addition to providing the bioanalyst the opportunity to contribute to the conversation by responding with comments about the draft guidance, becoming familiar with the expectations and requirements of guidances during their development helps the bioanalyst stay current and informed.

As with all fields of professional advancement and development, the Internet is a rich trove of information. There are some tremendously useful and effective online training courses dedicated to bioanalysis. These vary in quality, content, and accuracy, developed both by recognized and experienced opinion leaders and those with little tenure and few credentials. The convenience of online material is unquestioned and, as stated, some material is very helpful. Equally, some personal blogs, chat-rooms and other social media formats dedicated to bioanalysis can be excellent sources of current opinion and content. "User beware" applies here, though, and we recommend that new bioanalysts verify online-sourced information against peer-reviewed content. We expect that further advancement in online information and education will have a significant impact upon the future of bioanalysis. We encourage your participation with and support for the sources founded on accurate and helpful information.

1.3.2 Organizations and Focus Groups

There is now a multitude of nonprofit, professional organizations around the world that address bioanalysis areas. Regulated bioanalysis topics often feature heavily in the initiatives of these organizations as they seek to serve the interests of their members. Because these organizations are independent of pharmaceutical and biotechnology companies, these entities can often engage local health authorities on topics of bioanalysis. In recent years, groups such as the American Association of Pharmaceutical Scientists (AAPS), European Bioanalysis Forum (EBF), Japan Bioanalysis Forum (JBF), the Global Bioanalysis Consortium (GBC), and the Global CRO Council (GCC) have offered significant opportunities for bioanalysts to exchange ideas and interact with regulators. These and other professional organizations and their associated focus groups will continue to be a valuable asset to the bioanalysis community. We encourage bioanalysts to support and participate in the efforts of these groups. Opportunities to do so vary, from joining a local chapter to volunteering on programming and organization of meetings and conferences.

1.3.3 Conferences and Meetings

Through conferences, workshops, local meetings, and industry working groups, bioanalysts learn about and contribute opinions to the process of developing regulatory language, often interacting directly with representatives from regulatory bodies themselves. Many of the most popular and effective conferences are non-profit and conducted by the professional organizations referenced above. Others may have a for-profit structure but still offer valuable training and content. Rather than list all of the annual conferences with bioanalysis relevance, which are constantly changing in location and frequency, we encourage bioanalysts to participate in the community and follow its periodicals to learn of the regional, national, or international events worth attending.

Attendance at conferences and meetings is likely the most effective way for a bioanalyst to establish a personal bioanalysis network. As you will come to appreciate in the following chapters, information exchange is critical to a bioanalyst's professional development. The bioanalytical community welcomes new talent. As editors and years-long professionals engaging with this community, we can both attest to the value, collegiality, and enjoyment to be had working in this field. We hope you will join us in furthering the discipline and developing yourself as a professional bioanalytical scientist or supporter of the science so critical to bringing new drugs to the patients who need them.

References

1. Wagner JG. History of pharmacokinetics. Pharmacol Ther. 1981;12:537–62.
2. Hill H. Development of bioanalysis: a short history. Bioanalysis. 2009;1:3–7.
3. Shah VP, Midha KK, Dighe S, McGilveray IJ, Skelly JP, Yacobi A, et al. Analytical methods validation: bioavailability, bioequivalence and pharmacokinetic studies. Pharm Res. 1992;9:588–92. doi:10.1023/A:1015829422034.
4. Shah VP, Midha KK, Findlay JW, Hill HM, Hulse JD, McGilveray IJ, McKay G, Miller KJ, Patnaik RN, Powell ML, Tonelli A, Viswanathan CT, Yacobi A. Bioanalytical method validation—a revisit with a decade of progress. Pharm Res. 2000;17:1551–7. doi:10.1023/A:1007669411738.
5. USA Food and Drug Administration. Guidance for industry: bioanalytical method validation. MD, USA: Rockville; 2001.
6. Findlay JW, Smith WC, Lee JW, Nordblom GD, Das I, DeSilva BS, et al. Validation of immunoassays for bioanalysis: a pharmaceutical industry perspective. J Pharm Biomed Anal. 2000;21:1249–73.
7. DeSilva B, Smith W, Weiner R, Kelley M, Smolec J, Lee B, et al. Recommendations for the bioanalytical method validation of ligand-binding assays to support pharmacokinetic assessments of macromolecules. Pharm Res. 2003;20:1885–900.
8. Smolec J, DeSilva B, Smith W, Weiner R, Kelly M, Lee B, et al. Bioanalytical method validation for macromolecules in support of pharmacokinetic studies. Pharm Res. 2005;22 (1):425–1431.
9. Japan, MHLW, Guideline on Bioanalytical Method Validation in Pharmaceutical Development. 2013.
10. Japan, MHLW, Guideline on Bioanalytical Method (Ligand Binding Assay) Validation in Pharmaceutical Development. 2014.

Chapter 2
Understanding Bioanalysis Regulations

Mark Arnold, Eric Fluhler and Boris Gorovits

Abstract Bioanalytical data submitted to global health authorities in support of drug approval applications is expected to meet certain regulatory standards and requirements. Bioanalytical assay validation and the conduct of study sample analysis is addressed with a multitude of regulatory guidance documents. This chapter provides an overview of the regulations from a historical perspective to current day developments. Originally written around bioanalysis to support pharmacokinetic studies, bioanalysis regulatory language is now being applied to endogenous biomarker assays and evaluating immunogenicity of large molecule therapeutics. Quality management and documentation are also discussed in this chapter along with the process of how modern bioanalysis regulations are continuing to develop.

Keywords Bioanalysis · Regulatory · Regulated · LC-MS/MS · LBA · Guidance · PK · Immunogenicity · Biomarker

Charged with ensuring public health, health authorities enact regulations, guidance, and guidelines when they see deficiencies within the content of new drug filings or during inspections of facilities performing the work for such filings. The newly introduced regulations have ranged from minor to dramatic with many having a significant impact on the pharmaceutical industry practices. In many cases, new regulations address deficiencies observed in practices at individual companies but nonetheless are broadly extrapolated and applied; thus, many pay for the errors of a few. Over the past decade, there has been a steady expansion of not only the breadth

M. Arnold (✉)
5 Tulip Tree Drive, Burlington, NJ 08016, USA
e-mail: mark.arnoldcs@gmail.com

E. Fluhler
7 Cranberry Road, Washingtonville, NY, USA
e-mail: eric.fluhler@gmail.com

B. Gorovits
Biomedicine Design, Pfizer Inc., 1 Burtt Road, Andover, MA 01810, USA
e-mail: boris.gorovits@pfizer.com

M.L. Rocci Jr. and S. Lowes (eds.), *Regulated Bioanalysis: Fundamentals and Practice*, AAPS Advances in the Pharmaceutical Sciences Series 26,
DOI 10.1007/978-3-319-54802-9_2

of topics being regulated, but also the number of countries developing their own requirements. Due to this complex and sometimes conflicting set of rules, the pharmaceutical industry has responded with regional organizations and multiregional consortia providing feedback in an attempt to harmonize practices. This chapter provides a historical perspective on the issue, enabling the reader to understand the context from which today's complex bioanalytical regulations arose.

A historical timeline depicting the evolution of bioanalytical regulation and guidance is presented in Table 2.1. Good Laboratory Practices (GLPs) are the foundation upon which health authorities have built today's bioanalytical regulations. The first GLPs were enacted in 1972 in New Zealand and Denmark [1, 2]. In 1975, data irregularities found by the US Food and Drug Administration (FDA) in records at two laboratories resulted in congressional hearings. The hearings resulted in more inspection staff being added to the FDA, and a philosophical shift from trusting in the data submitted, to requiring demonstration that the data was reliable, with confirmation through agency audits. Subsequent to this, activities continued within the Agency culminating on December 22, 1978, when the GLPs became law within Section 21 of the Code of Federal Regulations Part 58 (21 CFR Part 58) [3]. Of note is that the US FDA GLPs only address animal safety studies. While widely followed for clinical bioanalytical studies, they are not required.

Since 1978, other countries and regions have implemented their own versions of the GLPs. In 1978, the Organisation for Economic Co-operation and Development's (OECD) Working Group released its "Principles of Good Laboratory Practice" [4] based on the 1976 draft FDA proposal on GLPs. The Principles were recommended by the OECD for use by member countries in 1981. Subsequently, a series of guidance documents for industry and health authorities were released, covering in detail additional aspects of GLPs, including the roles of the Sponsor and Study Director, the use of computerized systems, multisite studies and auditing, as well as updates to the primary GLP document [5].

The 34 OECD member countries follow the organization's GLPs and many nonmember countries also frequently follow them or similar local regulations. But, in an odd twist, while the FDA is a member country and has adopted many aspects of the OECD GLPs, unlike most of the OECD member nations, the FDA does not certify laboratories and facilities for GLP compliance. The FDA's inspection program is based on directed (inspectors sent to assess clinical (bioequivalence) and nonclinical [GLP] content within a filing) and nondirected (inspectors sent to GLP facilities that are frequently listed within filings) audits. While the outcome of these inspections may or may not result in the issuance of a Form 483 noting deficiencies in practice, the FDA does not issue a certification. This is increasingly becoming an issue for US sponsors since the UK's Medicines and Healthcare Products Regulatory Agency (MHRA) and the European Union's (EU) European Medicines Agency (EMA) have begun questioning the acceptability of clinical and nonclinical data from laboratories that have not had recent inspections [6].

In December 1990, industry and regulatory authorities representing the American Association of Pharmaceutical Scientists (AAPS), Federation International Pharmaceutique, Health Protection Branch (today known as Health Canada),

Table 2.1 Timeline of significant regulations impacting bioanalysis

Year	Regulation
1972	Denmark and New Zealand introduce GLPs
1978	USA: FDA GLPs Implemented, following 1977 identification of fraudulent records by FDA inspectors
1992	Canada: Health Canada—Guidance for Industry: conduct and analysis of bioavailability and bioequivalence studies—Part A: Oral dosage formulations used for systemic effects
1997	USA: FDA 21CFR Part 11 Electronic records; Electronic signatures
1997	Japan: Ministerial ordinance concerning the standards for the conduct of nonclinical studies on the safety of drugs (Ministry of Health and Welfare Ordinance No. 21, dated March 26, 1997)
1998	International: OECD Principles of good laboratory practice and compliance monitoring
2001	USA: FDA Guidance for industry: bioanalytical method validation
2003	USA: FDA Bioavailability and bioequivalence studies for orally administered drug products—general considerations
2003	Brazil: Resolution No. 899, guide for validation of analytical and bioanalytical methods
2005	Brazil: revision to 2003 ANVISA regulation
2005	India: Ministry of Health and Family, Guidelines for Bioavailability and Bioequivalence Studies
2005	International: ICH Q2(R1): validation of analytical procedures: text and methodology
2005	China: CFDA Technical guideline for human bioavailability and bioequivalence studies on chemical drug products
2009	International: WHO Good Clinical Laboratory Practice (GCLP)
2010	European Union: EMA guideline on the investigation of bioequivalence
2011	China: CFDA (2011) Guidance on management of laboratory for drug clinical trial biological sample analysis (interim)
2011	European Union: EMA Guideline on Bioanalytical Method Validation
2012	Brazil: ANVISA Resolution RDC 27, minimum requirements for bioanalytical method validation used in studies with the purpose of registration and post-registration of medicines
2012	Canada: HPFB conduct and analysis of comparative BA studies
2012	European Union: EMA Reflection paper for laboratories that perform the analysis or evaluation of clinical trial samples (GCLP)

(continued)

Table 2.1 (continued)

2013	Japan: MHLW Guideline on bioanalytical method validation in pharmaceutical development
2014	Japan: MHLW—Guideline on bioanalytical method (ligand-binding assay) validation in pharmaceutical development
2013	USA: FDA DRAFT Guidance for Industry: bioanalytical method validation
2015	European Union: EMA provides guide for reviewers that highlights the absence of US FDA laboratory certification process for GLP studies
2015	Canada: Health Canada requires stability testing to use 3 separate samples derived from separate containers, as opposed to 3 samples from a single storage container

US FDA, and the Association of Official Analytical Chemists met together at a conference entitled "Analytical Methods Validation: Bioavailability, Bioequivalence and Pharmacokinetic Studies." This was the first of what later came to be known as the Crystal City conferences, based on the location of this first meeting. The meetings objectives, which were met, included: (1) develop a consensus on required tests within a bioanalytical method validation and performance criteria; (2) how to apply those criteria to study sample analysis in bioavailability, bioequivalence, and pharmacokinetics studies; and [3] capture within a conference report all major consensus views for use in developing more specific regulatory guidance. The 3-day conference focused primarily on chromatographic methods (HPLC and GC with detectors that were available at the time, as LC-MS/MS was still very much in its infancy). The immunochemistry community was not forgotten at the meeting, but due to its lesser prevalence then as compared to today's post-biotechnology revolution, only a half-page of the conference report focused on immunochemistry techniques. Microbiological assays and kit assays were also discussed but given much less air time. The five-page conference report provided the consensus of the workshop and was viewed as the first step to not only harmonized practices across the industry, but also became a set of expectations against which both Canadian and US regulators began to assess drug concentration assays [7].

Moving much more quickly than its US counterpart, the Canadian Health Protection Branch released in 1992 its Guidance for Industry on the *Conduct and Analysis of Bioavailability and Bioequivalence Studies—Part A: Oral Dosage Formulations Used for Systemic Effects* [8].

That document used the concepts of the Crystal City meeting to describe the first regulatory language on bioanalysis and referred directly to the conference report and the GLPs:

> Bioavailability determinations rely on the adequacy of analytical methods for parent drugs and, when appropriate, their metabolites. This section describes the attributes of such methods and the validation procedures required in reports to establish and maintain selectivity, range, precision, and accuracy.

These concepts and the testing supporting them were delineated and incorporated into the Guidance in sufficient detail for their successful implementation by industry. The Guidance did not address which technology was covered, but given its limitation to orally administered drugs, it could be assumed that immunoassays for biologic drugs were not covered. However, as part of the required practices for bioavailability and bioequivalence study sample analysis, the fundamental concepts of today's incurred sample reanalysis (ISR) were defined when samples were analyzed in singlicate: "The purpose of the re-assaying is to establish that the degree of precision obtained with incurred samples is similar to that obtained for spiked standard or Quality Control (QC) samples." This requirement was dropped in 2003 [9], prior to the FDA-AAPS Crystal City IV discussion and the subsequent FDA expectation that ISR be routinely employed in a number of studies [10].

Another outcome of this regulation was setting the highest standard of performance for bioequivalence and bioavailability studies, studies central to labeling claims, which became more broadly applied to all nonclinical GLP studies and all clinical studies. In recent years, this broader application of very comprehensive and tough standards has been discussed frequently within scientific conferences and published on by both the industry and regulatory communities. While the concept of a tiered approach (lesser standards for some assays or studies and higher standards for others) exists, there has not been a consensus on which studies fall into each category. In this regard, some regulators have put forth the concept that studies affecting the prescribing label for a new drug should follow the highest standards. Thus, studies such as food effect, age–gender, and drug–drug interaction would be considered for the higher standard.

In 1997, the US FDA issued 21 CFR Part 11 governing the use of computerized systems, electronic records and electronic signatures [11]. This regulation, aimed at improving the use of computer systems, required that organizations operating under predicate rules (e.g., GLPs) ensured that the computer systems and resulting generated records also met the requirements of the predicate rules. For laboratories operating under the GLPs, structured computer system life cycle practices were adopted, electronic record maintenance and archive practices were improved, and audit trails were implemented to track the creation and editing of records in a GLP-compliant manner. The latter meant that during record generation, the "who did it", "when", and "what was created" needed to be tracked. In addition, when changes were made, the previous versions of the record needed to be maintained and the reason for the change documented and tracked. Even though there was no clinical counterpart to the GLPs, most bioanalytical labs used the same computer systems for clinical studies and achieved equivalent quality standards.

In December 1998, the US FDA released its draft "Guidance for Industry Bioanalytical Methods Validation for Human Studies" [12] for comment. Similar to the Canadian Guidance, the US FDA draft Guidance was focused on assay (1) accuracy, (2) precision, (3) sensitivity, (4) specificity, (5) linearity, and (6) reproducibility, but its simple Table of Contents belied the document's complexity over its Canadian predecessor. While titled for human clinical studies, it required that "The analytical laboratory conducting BA and BE studies should closely adhere to FDA's Good Laboratory Practices (GLPs) (21 CFR Part 58) and to sound principles of quality assurance throughout the testing process."

The industry response and discussions with the Agency resulted in the second Crystal City Workshop "Bioanalytical Methods Validation—A Revisit with a Decade of Progress." It was held in January 2000 and focused on chromatographic methods. Three months later, an AAPS workshop on macromolecule measurement was held. At both events, industry and health authority representatives actively discussed the science and regulations of bioanalysis. A combined report on both workshops was issued later in the year [13]. Having heard the scientific and practical concerns of the industry, the US FDA went on to finalize and release its final version of the "Guidance for Industry: Bioanalytical Method Validation" in May 2001. It became known as the BMV [14].

In 2003, Brazil's ANVISA released its bioanalytical regulations, which were updated in 2005 and more recently in 2012 [15]. Different still from the USA and Canadian Guidances, which are regulatory recommendations for best practices, ANVISA's regulations are law and leave no wiggle room for variations in science or compliance. Aimed primarily at BE studies, these regulations included specifics, such as using the simplest regression model and use of the "$T = 0$" measurement for assessing stability (i.e., using the actual concentration determined at the initiation of a stability study (time zero) as the comparator for subsequent stability time point measurements), that have generated challenges for industry practitioners. For HPLC or LC-MS/MS assays, use of any model other than linear required demonstrated statistical evidence that the chosen model was appropriate. Use of $T = 0$, rather than nominal concentrations for stability measurements in QC samples, meant that data demonstrating stability against nominal concentrations accepted by other countries could fail when compared to the $T = 0$ concentration data. This often required the reporting of stability data in two ways at a minimum for each method, and in some cases it required repeating stability measurements to obtain acceptable data for both approaches [10]. In later updates to its regulations, ANVISA harmonized its practices to assessing stability against nominal concentration and other updates have kept pace with changes in other regions.

Through two additional meetings with industry, Crystal City III (2005) and IV (2006), the FDA was able to refine a number of its expectations for bioanalysis and put forward the concept of incurred sample reproducibility as a means of assessing method robustness and analyst execution. The conference reports for these meetings, while not FDA Guidance, did become the basis for their expectations for bioanalytical method validation and sample analysis [16, 17]. Based on FDA observations of instances of poor data reproducibility in the field, the FDA's expectation for incurred sample reanalysis (ISR) was included in the Crystal City III consensus report. The requirement for repeating a portion of the samples to demonstrate assay quality was highly controversial with industry and Crystal City IV focused specifically on the issue, refining the concept beyond method quality to include a quality of execution assessment. ISR is now embedded within routine bioanalysis.

Further expansion of regional regulations came with the EMA issuing a draft and then finalized "Guideline on Bioanalytical Method Validation" in 2011. This Guideline provided a comprehensive set of requirements for chromatographic and ligand-binding immunoassays based on the then current state of the art for both technologies [18]. It became the de facto global standard for bioanalytical regulation. The Guideline included a few items that further harmonized practices, but for the most part industry was already in compliance. A larger, but less noticeable change came with a 2012 EMA reflection paper of Good Clinical Laboratory Practices (GCLP), which blended aspects of Good Clinical Practice (patient confidentiality and safety) with documentation and evidentiary aspects of the GLPs, thus requiring all bioanalytical clinical and nonclinical work to have identical compliance standards [19]. While most of the industry already analyzed clinical samples in accordance with the principles of GLP or in a GLP-like manner, this reflection paper put forward the equivalent standard and eliminated any uncertainty in practice.

Regional expansion continued in 2012 when Health Canada indicated that it would operate under the EMA's BMV Guideline [20]. In 2013 and 2014, Japan's Ministry of Health, Labour and Welfare (MHLW) issued its regulations for bioanalysis using chromatographic and ligand-binding assays, respectively [21, 22]. These guidances were consistent with the EMA Guideline, but were less strict and provided more flexibility to incorporate science into bioanalytical practice. They also differed from the FDA and EMA guidance/guideline creation process in that MHLW's working group included industry representatives as part of the writing team, rather than relying on industry comments to draft documents. In 2013, the FDA issued a draft revision to its BMV guideline [23] that resulted in both positive and negative responses from industry, and another AAPS-FDA Workshop, Crystal City V. The workshop report included only minor changes to drug concentration measurement practices. While that workshop report did provide an initial framework on the FDA's inclusion of biomarkers by chromatographic and ligand-binding assay (LBA) methods within the BMV, no specific recommendations were made. An in-depth discussion of the unique issues related to measuring biomarkers was continued with the FDA at Crystal City VI in 2015. At the time of the writing of this chapter, its conference report is pending. Similarly, the final revision to the FDA's revised BMV guidance, as well as an English version of a guidance from China's CFDA, are also pending.

2.1 Industry Collaboration

For more than 25 years, the pharmaceutical industry's bioanalytical community has communicated, collaborated, and worked with various health authorities to implement science-based regulations. In addition to the AAPS Crystal City meetings, others organizations (EBF, APA, CFABS) have held conferences, some with health authority representatives, and produced conference reports/white papers that reflect consensus positions held by conference attendees. However, for most of that time, the content of these conferences tended to be region specific. That changed in 2010, when bioanalytical scientists from Canada, the EU, and USA joined together to form the Global Bioanalysis Consortium (GBC). Representing a number of regional organizations (Applied Pharmaceutical Analysis, AAPS, Canadian LC-MS/MS Group, and European Bioanalysis Forum), this group invited other regional organizations to join or to form groups and join. Thus, organizations in India (APA-India) and Brazil (AC-Bio) joined, and organizations formed and joined from China and Japan (China Bioanalysis Forum and Japan Bioanalysis Forum). As a worldwide organization consisting of representatives from various scientific associations, the mission of the GBC was "to harmonize and merge existing or emerging bioanalytical guidance to create one, unified consensus document that can be presented to the regulatory bodies/health authorities in various countries" [24]. This mission was fulfilled through the formation of global working groups charged with developing scientific best practices for bioanalytical chromatographic and immunoassay techniques. The fruits of their work resulted in a number of consensus white papers. These white papers are a valuable resource in understanding the science behind regulatory expectations [25].

2.2 Bioanalytical Method Validation and Sample Analysis in Support of Pharmacokinetic Studies

As mentioned previously, early discussions on the content of BMVs for drug development emphasized methods used for the measurement of drugs and metabolites in biological fluids as important components of pivotal bioavailability and bioequivalence studies [8]. Over time, however, the conduct of BMV has become a standard expectation for a much broader range of drug development studies. Indeed, the 2012 EMA guidance on BMV, which is amongst the more comprehensive and contemporary guidance at the time of this printing, indicates it is applicable to "animal toxicokinetic studies and all phases of clinical trials" [18]. Hence, in practice companies often initiate the first bioanalytical method validation on a drug candidate in preparation for the first GLP toxicology study, which is typically well in advance of any regulatory submission or clinical trial. Subsequent to these first in human (FIH) enabling toxicology studies, BMV in human matrix is undertaken in preparation for the FIH study. Generally, a "full" BMV is performed on the parent drug and any major metabolites [26, 27] in each species and matrix used during the conduct of regulated (GLP or GCP/GCLP) studies. Supplemental or "partial" validations are conducted when changes are made to a previously validated method. Typical alterations accommodated through a partial validation include changes in the following: anticoagulant type, storage conditions, assay range, equipment, matrix (e.g., serum vs plasma), reagents (e.g., critical ligand binding assay reagents), laboratory conducting assay, population (e.g., healthy volunteers versus patients), and concomitant medications. Tests conducted during partial validation focus on demonstrating the validity of aspects of the method likely to be impacted by the change in assay conditions. Hence, a partial validation can vary from as little as a single analytical run to almost a complete revalidation, depending on the nature of the change. Cross validation is a supplemental validation conducted to compare results when two distinct methods are used either within a study or across studies during the development of a compound. Cross validation should also be considered when sample analyses within a study are analyzed at more than one site/laboratory. The goal of a "cross" validation is to demonstrate intermethod or interlaboratory consistency and reliability.

The most comprehensive and widely referenced regulatory guidance on BMV are those from the EMA (effective in 2012, [18]), Japan's MHLW (effective in 2013 and 2015 for LC-MS/MS [21] and LBA [22] methods, respectively), Brazil's ANVISA (effective in 2012, [15]) and the U.S. FDA (effective 2001, [14]). As mentioned previously, the FDA is in the process of updating its 2001 Guidance to reflect many of the outcomes of the Crystal City meetings which have taken place since its issuance. Since there are subtle differences in validation requirements among guidances, scientists performing contemporary BMV for international submissions must consider all of them to develop an inclusive validation and analysis package. This being said, all guidance documents require demonstration of the following fundamental method attributes within certain criteria (unless otherwise justified in the method and validation documentation):

- Selectivity—the ability of the method to measure the analyte in the presence of other sample components.
- Accuracy—the closeness of test results from the method to the true (or nominal) value (concentration).
- Precision—the variability between individual test results determined in repeated measurements.
- Sensitivity or Lower Limit of Quantitation (LLOQ)—the lowest concentration of analyte which can be quantitated reliably.
- Calibration Curve—the relationship between analyte concentration and instrument response (also referred to as Standard Curve).
- Stability—the stability of the analyte in the matrices, containers and conditions used for sample collection, storage and analysis. Note: Health Canada has recently released specific requirements for the use of at least 3 replicate tubes exposed to each stability condition (each time point, condition, etc.) [28].
- Dilution Integrity (when applicable)—the ability to generate accurate and precise results when diluting a high concentration sample into the calibration range of the method.
- Carryover (when applicable)—when the potential for carryover exists, it should be investigated by measuring the analyte response in blank samples placed immediately following a high concentration sample.
- Matrix Effect—the alteration of analyte response due to the presence of other matrix components.

- Reproducibility—the precision of the method under the same operating conditions over a short period of time (several runs or days). This is typically determined using matrix spiked QCs. Also see "Incurred Sample Reproducibility" below.
- Reference Standards—analytical standards of known identity and purity should be used during method validation and sample analysis. A certificate of analysis or other documentation supporting the identity and purity of the standard should be available for each batch used.

For specifics on the acceptance criteria required for each of these validation elements, the reader is directed to the guidance documents themselves.

In addition to providing specifics on the conduct of method validation, the aforementioned BMV guidance also describe run acceptance criteria and quality expectations for the conduct of routine sample analysis. Key expectations are described for the following:

- Run Acceptance—Criteria on the use (e.g., number and concentration) and performance of blank samples, calibration curve standards and quality control samples included within the run.
- Reanalysis of Study Samples—Instances when it is appropriate, process and documentation needed for reanalysis of study samples. This type of reanalysis can be broken down into two categories: (1) Reanalysis for analytical reasons, where there is an assignable cause such as instrument failure, documented dilution error, poor chromatography, incongruence of replicate results in LBAs, failure of run acceptance criteria, concentration above the Upper Limit of Quantitation (ULOQ), etc.; (2) Reanalysis for pharmacokinetic (PK) reasons (incongruous result) such as having measureable drug in pre-dose, placebo, or control samples, or where a sample switch seems likely. Note, some guidances prohibit and most strongly discourage reanalysis for PK reasons during the conduct of BE studies; in any situation these reanalyses should be guided by a pertinent SOP that clearly defines circumstances in which sample reanalysis is permissible, the sample reanalysis procedure, and the manner in which a final reported result is determined from all replicate measurements that have been made. When reanalyzing samples, the laboratory is expected to provide a clear justification for why a reanalysis was performed, and the manner in which a final reportable result was generated [18].
- Incurred Sample Reanalysis (ISR)—ISR is broadly defined as the selection, analysis, and reporting of a subset of study samples; reanalyzed to establish the reproducibility of the method using incurred samples. ISR is generally performed once per species used in toxicokinetic assessments and as part of pivotal clinical studies. The ISR process is now viewed as a required and critical element of bioanalytical method assessment. The main purpose of ISR is to ensure reliability in reported sample concentrations. Because ISR investigation requires access to incurred study samples, the actual test occurs during the sample analysis phase and therefore reflects both the validity of the assay and its

execution. ISR may also provide information related to the stability of the analyte(s) under sample storage conditions for the time period between initial and ISR testing. Together with data generated by analysis of reference material spiked QC samples, ISR data demonstrates precision and accuracy characteristics for a given bioanalytical procedure. During ISR analysis, the original and repeat sample analysis results obtained by the same method but in different runs are compared. The requested number of samples to be tested in ISR analysis varies among regulatory guidelines.

ISR should be established at least once for each method in nonclinical species. Requirements for ISR investigations in human studies vary between regulatory guidelines but are generally expected for BE and all pivotal PK/PD studies, notably those whose results will be incorporated within the prescribing label.

Examples of clinical evaluations where ISR is recommended are:

- First-in-human studies
- First-in-patient studies
- Drug–drug interaction studies
- Studies in hepatic or renally impaired populations
- BE and biocomparability studies
- Pivotal Phase III studies
- First time use of a method in a new laboratory.

Results of an ISR evaluation may be reported as an addendum to the assay validation report or as part of the sample analysis report for a study. Investigations into possible causes of ISR failures should be carefully documented to allow for detailed study reconstruction.

Again the reader is referred to the individual guidance documents for specifics on each of the above topics related to routine sample analysis for pharmacokinetic assessments.

2.3 Validation and Application of Assays for Immunogenicity Assessments

The majority of protein therapeutics are known for their potential to induce a drug-specific immune response that is commonly evaluated by the measurement of antidrug antibodies (ADA). Treatment-induced ADA response is expected to vary in immunoglobulin isotype (e.g., IgG, IgM, IgE), affinity and epitope specificity of drug binding and may include drug neutralizing antibodies (NAb). NAb can be defined as an ADA that is capable of completely or partially blocking drug interaction with its target. Induction of drug-specific ADA often results in various degrees of impact on drug pharmacokinetics (PK) including shorter drug half-life due to faster clearance of large drug–ADA immune complexes. A direct impact of drug-specific ADA on the performance of a PK assay is also possible as the ADAs may interfere by blocking the drug analyte binding with assay reagents. For example, ADA with neutralizing specificity may block drug-target binding, therefore impacting PK assay performance if target or a molecule with similar specificity of binding is used as one of the assay reagents. A careful investigation may be required in order to determine the true nature of ADA impact on the observed PK profile. Detection and evaluation of ADA development, particularly in clinical studies, is a critical element of biotherapeutic development. Initial groundwork toward the harmonization of ADA-detecting assays was made by industry groups [29] with the focus on developing recommendations for the design of ADA-detecting assays. Similarly, industry groups pioneered steps and methodologies required for the development of assays set up to detect the presence of ADAs with drug neutralizing activity (NAb) [30]. Particular focus was directed toward developing risk-based strategies for ADA evaluation and characterization [31]. Later, industry positions on ADA and NAb assay development were presented [32]. The EMA position was finalized and incorporated in the EMA Guideline on Immunogenicity Assessment [33], followed by a draft FDA Guidance for the Industry [34]. At the time this chapter was written, the FDA guideline remains a draft document. The above two regulatory guidance focus on the approaches and various critical aspects of importance during ADA and NAb detecting method development and validation. In addition, FDA has issued a regulatory guideline designed to provide recommendations for immunogenicity risk factor identification and mitigation [31]. This 2014 FDA guideline provides a critical connection between product immunogenicity risk assessment and design of appropriate immunogenicity testing strategies. Most recently, the EMA has issued a draft update of its guideline [35].

Typically, a tiered approach to immunogenicity evaluation is applied and includes an initial screening test followed by an ADA specificity confirmatory test. In addition, tests may be conducted to determine final reportable ADA titer value, characterize ADA isotype, and to determine whether the ADA has a drug neutralizing specificity (NAb). Drug-specific immune response is expected to be polyclonal, i.e., to contain a mixture of various isotype immunoglobulins with different epitope specificity and binding affinity to the drug. The initial ADA assay is expected to detect ADAs with all possible specificities, affinities, and isotypes. Therefore, it is not possible to generate a true reference material that would be able to reflect performance of the variety of analytes expected to be found in the sample. Positive Control (PC) samples are routinely used to monitor ADA assay performance, but do not allow for a quantitative reporting of ADA assay results. Consequently, ADA assays most typically report semiquantitative results, (e.g., titer values) [36]. The same can be said about neutralizing antibody (NAb) assays. While ADA assessments usually employ ligand binding assays (LBA), NAb assays can be cell or noncell based, depending on the drug's mode of action (MOA). NAb assays are most applicable during clinical investigations and are rarely conducted in support of nonclinical studies, unless a translation between nonclinical and clinical safety signals can be drawn.

Details regarding ADA assay development and validation can be found in several regulatory guidelines and industry white papers [29, 34–36]. In the initial screening test, samples are identified as putatively positive versus negative based on comparison with the screening assay cut point value. The cut point value is developed during ADA assay validation by statistically analyzing the distribution of signals generated from the analysis of individual samples obtained from drug naïve subjects representative of the study population. Typically, the 95th percentile of the distribution is compared to the performance of the assay negative control to determine an assay-specific cut point correction factor. The particular distribution cutoff is chosen to ensure that all putative positive samples are detected by establishing a 5% false positive rate in the initial screening assay investigation. Putative ADA positive samples are then confirmed for the specificity of binding to the drug by, for example, analysis of the unlabeled drug's ability to inhibit signal in the ADA assay. Here again, ADA—drug specificity of binding is confirmed through comparison to a predefined, assay-specific confirmatory cut point. As for the screening assay, the confirmatory assay cut point is defined through statistical analysis of signal inhibition values generated by drug naïve, study population specific samples. A sample is defined as ADA positive and drug-specific if the corresponding screening signal is equal to or greater than the screening assay cut point value and the percentage inhibition value generated in the confirmatory assay is equal to or greater than the confirmatory assay cut point value.

During sample testing, positive and negative control samples are routinely applied to determine the appropriateness of assay performance based on predetermined acceptance criteria. Assay Negative Control (NC) typically consists of a pooled drug naïve, study-relevant, species matrix sample. Assay PC is typically a matrix sample spiked with either a monoclonal or polyclonal, drug-specific, non-human origin antibody raised by hyperimmunizing various laboratory species. Several of the assay characteristics are determined based on the PC material performance, including assay sensitivity and drug tolerance. The regulatory expectation is that a clinical ADA assay should be able to detect 250 ng/mL of PC material (500 ng/mL for nonclinical assays). Assay drug tolerance is a parameter that reflects the ability of unlabeled drug to block ADA binding to the assay reagents (e.g., biotin-labeled drug) and, therefore, inhibit ADA assay signal. Residual drug found in the sample can subsequently be expected to reduce ADA assay sensitivity. The ADA assay ability to tolerate a certain amount of drug in the sample can be assessed based on the performance of PC material, although this does not directly reflect sample ADA performance. Numerous variations of ADA protocols intended to improve assay tolerance have been reported [37, 38].

Other parameters evaluated during ADA assay development and validation include:

- Identification of appropriate assay reagents, selection of the assay PC
- Identification of appropriate NC for the assay
- Selection of assay analytical platform and format

- Establishment of assay conditions, including assay buffer, reagent conjugation conditions, minimum required dilution (MRD), assay incubation times
- Identification of confirmatory assay conditions including drug concentration spiked
- Determination of both Screening and Confirmatory cut point values during ADA assay validation
- Determination of assay precision, robustness and sensitivity using assay PC material performance
- Establishment of acceptance criteria based on PC and NC material performance.

Study incurred samples are commonly tested at multiple dilutions to produce a final reportable titer value. Titer value can be determined as the maximum sample dilution that produces signal at or above the assay cut point value. Similar approaches to report NAb assay data are also applied.

2.4 Chromatographic Versus Ligand-Binding Assays

Compared to LC-MS/MS assays, typical LBA methods do not include a significant sample pretreatment step. Neat matrix samples are usually diluted in order to fall into the assay range of quantitation and if a higher than the minimum-required dilution (MRD) is required, assay diluent buffer containing related matrix is used. Matrix components can play a significant role in the final assay performance based on their ability to interact with the analyte, various assay reagents and components (e.g., plates or beads). The concentration and binding properties of these non-specific and generally low-affinity interferences frequently vary from patient to patient and are highly dependent on the nature of assay matrix, disease, and sample collection strategy. The aspects of selectivity and specificity are, therefore, critical to LBA methods. Matrix component interference can be particularly noticeable at the lower and upper limits of the detection curve. As a result, it is critical to understand assay performance at the LLOQ/ULOQ levels [18].

Biotherapeutics may be applied at high doses while LBA methods typically have a narrow dynamic range of quantitation. Therefore, samples can be diluted 100,000 fold and higher. A careful evaluation of the dilutional linearity of LBA assays is critical to ensure accurate reporting of sample concentrations. Dilutional linearity is expected to be performed during assay validation and should demonstrate an assay's ability to detect accurately the analyte concentrations similar to expected Cmax levels. This test is particularly important as LBAs commonly produce non-linear sigmoidal response curves which are analyzed using four- and five-parameter logistics or similar fitting algorithms.

A parallelism analysis, in principle, is a test that is similar to dilutional linearity but is designed to address questions related to possible analyte modifications while in circulation in vivo, including partial degradation or other forms of covalent or non-covalent modifications.

Relatively high well-to-well variability observed for older generation, plate-based LBA protocols resulted in the requirement to conduct duplicate measurements per sample. Similarly, acceptable ranges for assay precision (20%) and accuracy (25% for LLOQ/20% for other QCs) have also been extended compared to a typical LC-MS/MS method. A requirement for duplicate analysis has been recently challenged and may evolve with the advent of newer and more precise LBA platforms. The nonlinear nature of a typical LBA concentration–response curve often calls for inclusion of standards, known as anchor points, which extend outside of the quantifiable range. Inclusion of anchor points may improve mathematical fitting of the calibration curve and, therefore, improve QC and standard performance. At least eight calibrators with at least a 75% passing rate are expected to be included in the assay range of quantitation in order to support robust fitting of the data.

2.5 Biomarkers

Biomarker measurements are typically performed to reflect a drug's pharmacodynamic (PD) effects. While the requirements for PK and ADA analysis are well described in the existing regulatory guidance, regulatory expectations for biomarker analysis are much less mature. In fact, the FDA's 2013 draft guidance is the only BMV guidance that mentions biomarkers. As previously discussed, even the most recent Crystal City VI meeting offered only a preliminary and high-level framework for biomarker validation requirements. This cautious pursuit of biomarker bioanalysis regulations by both regulatory and industry scientists is clear recognition of the unique challenges biomarker analysis presents, and indicates a need for the science and practice related to biomarkers to mature before being codified within regulations. Among the unique aspects of biomarkers are the endogenous nature of the analyte and the difficulty in producing a reference standard that is identical to the endogenous compound (e.g., proteins). Other factors are the complex matrix in which many biomarkers reside (e.g., membranes, cells, tissues) and the rapid introduction of new and innovative approaches for measuring biomarkers. These challenges make it difficult to settle on a stable set of quality standards that can be broadly applied.

2.6 Other Laboratory Tests

The above regulations apply to the work typically conducted in bioanalytical laboratories as part of PK, immunogenicity, and PD assessments during drug development. It is important to recognize that there are a large number of laboratory tests performed during drug development and after drug approval that fall outside the bioanalytical realm. These tests may be conducted for a variety of reasons that include, but are not limited to: patient diagnosis, drug safety testing, clinical pathology, or even therapeutic drug monitoring. These tests are beyond the scope of traditional bioanalysis and hence this book, and fall under their own set of regulatory requirements. A few of these from the USA are described briefly below to add context for the reader. It should be noted that similar practices and rules exist in other regions.

- Clinical Laboratory Improvement Amendments (CLIA): a quality/compliance program directed by the Division of Laboratory Service within the Centers for Medicare and Medicaid Service in the United States. The CLIA program regulates laboratories performing tests on patient samples to ensure they produce accurate and reliable test results. It is noteworthy that CLIA does not address the validity of an assay, that is, its ability to measure or predict the presence or absence of a clinical condition, which falls under FDA jurisdiction. CLIA laboratories must establish and maintain certain performance characteristics on their assays and must pass state administered inspections on a biennial basis [39].
- Laboratory Developed Test (LDT): A LDT is a type of in vitro diagnostic device that is designed, manufactured, and used within a single laboratory. These tests typically identify, detect, or measure the presence or absence of a clinical condition or predisposition in a patient. In 2014, the FDA released its first draft framework for the oversight of LDTs, citing the risk that flawed LDTs may pose to patients [40]. The FDA also cited a gap in current regulations with respect to their oversight of LDT validity (ensuring they are safe and effective). In the draft framework, FDA describes a three-tier, risk-based system that it will apply to regulate LDTs beginning six months after the final guidance is issued.
- Premarket Approval (PMA) and 501(k): Many commercial devices used to perform clinical tests on patient samples require regulatory approval prior to marketing and use. The type of application [PMA or 510(k)] depends on the associated risk to the patient and whether a test already exists for the endpoint. FDA review of these devices (tests) is focused on ensuring they are safe and effective for their intended use [41].

2.7 Quality Management in the Regulated Bioanalytical Laboratory

In addition to the specific requirements outlined in various BMV guidelines, laboratories conducting regulated bioanalysis are also expected to establish a robust quality management system (QMS). For laboratories conducting regulated, non-clinical bioanalysis, the fundamental expectations of a QMS system are described within the FDA and OECD GLPs [3, 4]. Similar expectations exist for laboratories conducting clinical bioanalysis, and a contemporary guidance on GCLPs is available in the previously mentioned EMA reflection paper on the topic [19]. The GLPs and GCLPs (collectively known as "GxP") provide a framework for the facilities, systems, procedures and documentation that should be present to ensure the reliability, quality, and integrity of the work performed and associated results. Key components of a contemporary QMS supporting GxP bioanalysis include, but are not limited to:

- Organization, Personnel, and Facilities—Management of establishments conducting GxP work should ensure staff roles and responsibilities are established and documented prior to conducting regulated work. Staff engaged in regulated study support should have the appropriate education, training, and experience to conduct the assigned work and this information should be documented and updated regularly. Test facilities should be of suitable size and construction to meet the requirements of the study. This includes an adequate degree of separation between activities to prevent adverse effects on the study (e.g., contamination within or between studies). Facilities should also have an appropriate controlled access space for storing study specimens and archiving data and reports in an environment that prevents deterioration.
- Standard Operating Procedures (SOP)—Testing facilities should have written SOPs that are approved by management and ensure the quality and integrity of the data generated in the course of the study. Deviations from SOPs related to the study should be acknowledged by the study director (nonclinical) or sponsor representative (clinical). A controlled and documented SOP revision process is required. SOPs related to bioanalysis may cover material and reagent preparation and handling, equipment operations and maintenance, receipt and storage of study samples, protocols relevant to assay validation, sample testing, records keeping, training and QC processes. Method-specific documentation sometimes is also issued as a method-specific SOP. Laboratory staff are expected to maintain role-appropriate levels of documented training as defined in relevant SOPs.
- Quality Assurance Program—Test facilities should have a Quality Assurance (QA) program to assure the studies performed are in compliance with GxPs. The QA program should be carried out by individuals designated by management and not involved in the conduct of the study. The QA unit should inspect each GxP study at intervals adequate to assure the integrity of the study and should maintain proper documentation of each inspection. Different levels of QA review can be conducted, including facilities, systems, equipment, personnel training, data recording, and reporting audits. Inspections should also determine that protocols and SOPs have been made available to study personnel and are being followed. Any issues found during the course of inspection that are likely to affect study integrity should be brought to the attention of the study director (nonclinical), sponsor representative (clinical), and laboratory management in a timely manner.

- Equipment, Materials and Reagents—Equipment, including validated computerized systems used in the generation, storage, and retrieval of data, and for environmental control of factors relevant to the study, should be suitably located, of appropriate design and capacity, and properly maintained. Chemicals, reagents, and solutions should be labeled to indicate identity, concentration, storage requirements, and expiration date. Information on the source, preparation date, and stability should be available.
- Sample labeling, storage and chain of custody—Samples should be labeled in a manner to allow unequivocal identification. A mechanism to ensure that samples were maintained under appropriate storage conditions prior to analysis (shipping, storage, etc.) should be implemented and maintained. This will typically include a tracking system that provides chain of custody from receipt through destruction.
- Patient Safety, Informed Consent, and Study Blind—In the case of clinical bioanalysis, the safety of the trial patients or subjects take precedence over all other aspects of the study. Therefore, lines of communication must be present between the laboratory, sponsor (or their representative), and clinical investigators to ensure that any issues that impact patient/subject safety are communicated rapidly. There should also be a mechanism to ensure that the laboratory is informed of what actions to take if informed consent is withdrawn by a patient or subject. Many clinical trials are blinded in order to ensure an unbiased approach in interpretation of trial data, including information on treatment outcome and adverse effects. For blinded clinical trials, laboratories should exercise due diligence to ensure they do not inadvertently compromise the study blind while communicating with investigators. In cases where bioanalytical data could unblind the trial, particular care must be taken to ensure data is only communicated with an established point of contact. A double-blinding approach is most typical during conduct of a drug development trial [18]. In a double-blind trial design, the subjects, as well as the investigator and staff involved in administering the treatment, are not aware of the received treatment. Bioanalytical scientists may be unblinded prior to sample analysis or after review of sample concentration data, but procedures must ensure the unblinding is restricted to appropriate personnel and does not impact the trial design. A set of SOPs is expected to prevent inadvertent and inappropriate distribution of treatment information, particularly during reporting of the data, to the staff that is expected to remain blinded for the duration of the study

2.8 Documentation

Thorough documentation of bioanalytical activities is a central and common expectation for all regulated work. All major regulatory guidelines are generally in high agreement regarding assay documentation expectations [14, 18]. The level of detail included in the documentation should allow for reconstruction of the study as it was conducted. Reports summarizing results of activities conducted during assay validation and sample testing are expected at the conclusion of each study.

While validation activities are not formally required to be conducted under GxPs, in practice most laboratories conduct validations using the same practices and procedures applied during regulated sample analysis (in some cases excluding the QA audit activities). The main goal of documenting assay validation data is to retain and record the performance characteristics of a successfully validated method in order to demonstrate its suitability and reliability for the intended application. During the course of bioanalytical study support, assays may require additional validation or revalidation. For example, a change in the assay reagent lot, assay matrix, or disease indication may require additional tests to ensure that the same or modified assay continues to meet predefined requirements. These and similar experiments should be carefully documented in an updated version of the assay procedure and validation report. When reporting validation data, all run information should be disclosed, including data from failed runs.

Bioanalytical data generated during the application of an assay for study sample analysis is commonly documented either in an independent bioanalytical report or a combined bioanalytical/PK report. Assay investigations conducted during study support are expected to accompany the study report and include information on issue resolution and any required assay modifications.

Typically, the following information is expected to be included in the assay documentation package maintained at the lab:

- Source (or raw) data.
- Summary of assay performance.
- Assay procedure including information on assay components and required steps.
- Assay validation report(s).
- Study sample analysis report(s).
- Failure investigations, deviations, assay changes and modifications including all supporting information.
- Assay relevant correspondence records.
- References or copies of relevant SOPs.
- Records identifying appropriate qualification, training and experience of assay personnel.
- Instrumentation maintenance and calibration records.

Documentation can be stored in either paper or electronic format, and is expected to record events in a contemporaneous manner. Although printing electronic records and storage of the paper version with deletion of the electronic record was once acceptable, most agencies currently oppose this practice. Appropriate laboratory procedures and policies should be in place explaining documentation generation and storage, including conversion of paper documents into an electronic format. The latter needs to include a thorough QC process to ensure electronic copy completeness and equivalence to the original paper copy. Any clarifying modifications to the documentation added later should not obstruct previously created information and should be accompanied by the reason for the change. If stored in electronic format, relevant regulatory guidance is applicable, including existence of an audit trail.

Details related to the information to be documented as part of the assay package are thoroughly described in regulatory guidelines (FDA, EMA and MHLW) and GBC Team A8 report [42].

2.9 Conclusions

As technologies advance alongside new and evolving drug molecules and delivery systems (e.g., sRNA, adnectins, nanobodies, millamolecules, viral vector delivery systems), the nature of a bioanalytical assay will also continue to change. As observed by regulators over the past 25 years, these types of advances drive re-evaluation and updating of the performance characteristics expected of the assays during validation and sample analysis. In this way, regulators are able to ensure the quality of the data generated, and consequently the safety and efficacy assessments of new therapies.

In the realm of LC-MS/MS, small molecule bioanalysis has been advanced with UHPLC, low flow systems, newer more sensitive instruments, high resolution accurate mass instruments, and full scan systems of improved sensitivity. UHPLC has increased the throughput of assays without the loss of chromatographic resolution. Micro- and nanoflow systems are increasing sensitivity. The increases in sensitivity introduced by triple stage quadrupole mass spectrometers have kept pace with the development of more potent molecules, and thus continue to provide valid pharmacokinetic assessments. High resolution instruments with improved specificity have ensured that fewer interferences plague bioanalytical scientists and also enable the bioanalytical support of microdosing studies that use smaller numbers of patients. The use of full scan instruments and targeted selected reaction monitoring is enabling early identification and quantitation of metabolites with the practices for validation and sample analysis being worked out based on current practices for triple stage quadrupole mass spectrometers. Combined, these advances provide a larger variety of tools for bioanalytical scientists and enable more reliable and sensitive assays.

The diversity of therapeutic modalities, particularly in the field of biotherapeutics, continues to evolve. Alternative scaffold antibody and fusion proteins, bispecific antibodies, antibody–drug conjugates, nucleotide and viral vector-based biologics are now a part of the clinical development space. Such second generation biotherapeutics present additional bioanalytical challenges including increased requirements for assay sensitivity, specificity, the ability to handle smaller sample volumes, larger dynamic range of the assay, and measure specific analyte type (e.g., free versus total monoclonal antibody moiety). A new generation of LBA methodologies has advanced to address some or all of these requirements. Many of the new LBA analytical technology platforms have moved away from a plate-based approach and may allow testing more than a typical 20 to 30 samples per analytical run. This raises a question of whether a typical LBA assay run should contain a single set of standards and QCs or should incorporate additional QCs throughout the run set. Currently, it is generally agreed that an analytical run can be defined as a set of standards with interspersed QCs and that a number of QCs greater or equal to five percent of the total number of unknown samples should be present in a run to appropriately control assay quality [43]. Questions related to the detailed positioning of QCs within the run, and size of a run batch are still left open. Currently, in a majority of LBA methods, samples are tested in duplicate; while improved precision for some LBA-based analytical platforms may allow for a single replicate analysis.

Affinity capture (AC) LC-MS/MS technology is quickly being adopted for analysis of biotherapeutics. A good example of an AC-LC-MS/MS application is the analysis of various drug species present following the administration of an antibody–drug conjugate (ADC). AC-LC-MS/MS methods are a hybrid between LBA and LC-MS/MS platforms as they include an initial affinity capture step (e.g., by using specific antidrug antibody) followed by a digestion and peptide analysis on an LC-MS/MS platform. The LC-MS/MS step adds improved specificity of detection while the AC step provides an opportunity to improve assay selectivity and sensitivity. Importantly, the existence of an AC step brings challenges similar to those observed for a typical LBA method, including potential matrix component interference, impact of sample pretreatment (e.g., dilution) on the analyte type (e.g., free/total), and nonlinearity of response. Hybrid assays may, therefore, have inherently higher variability versus other types of LC-MS/MS methods. The overall experience for both regulatory agencies and the industry with the AC-LC-MS/MS platform is still developing. Currently, there is no guideline that clarifies regulatory expectations for AC-LC-MS/MS assays.

Because of the long duration of a typical drug development cycle, bioanalytical methods and analytical platforms applied early on may be reassessed and changed later in development. An improved understanding of assay sensitivity or the type of analyte requirements may drive a platform change, for example requiring a switch between LBA platforms or a change from LBA to an AC-LC-MS/MS platform. Regulatory agencies generally expect presentation of data that support and justify the implemented analytical platform change. Some of the technologies are sufficiently different to prevent a direct comparison of absolute analyte concentrations. Currently, consensus exists that a set of bridging data should be presented to directly compare different analytical methods.

Details on current scientific practice will be discussed within other chapters, with this chapter having established the historic precedents and progress leading to today's extensive and sometimes conflicting set of regulations. As this chapter goes to press, the International Council on Harmonization announced acceptance of a proposal to generate M10—a harmonized approach to bioanalytical method validation and sample analysis. While this effort may take some time to come to fruition, it is a positive step forward for regulators and the scientific community.

References

1. New Zealand, Testing Laboratory Registration Act. 1972. Ministry of Research Science and Technology.
2. Denmark, Danish National Testing Board Act No. 144. 1973.
3. USA, Good Laboratory Practice for Nonclinical Laboratory Studies. 1978. Code of Federal Regulations Part 58.
4. Organisation for Economic Co-operation and Development Working Group, Principles of Good Laboratory Practice. 1978. Available from: http://www.oecd.org/officialdocuments/publicdisplaydocumentpdf/?cote=env/mc/chem(98)17&doclanguage=en.
5. OECD Series on Principlies of Good Laboratory Practice and Compliance Monitoring, N 1. 1998. Available from: http://www.oecd.org/chemicalsafety/testing/oecdseriesonprinciplesofgoodlaboratorypracticeglpandcompliancemonitoring.htm.
6. European Medicines Agency, Triggers for audits of Good Laboratory Practice (GLP) studies. 2015.
7. Shah VP, Behl CR, Flynn GL, Higuchi WI, Schaefer H. Principles and criteria in the development and optimization of topical therapeutic products. Pharm Res. 1992;9(8):1107–11.
8. Canadian Health Protection Branch. Conduct and analysis of bioavailability and bioequivalence studies—Part A: oral dosage formulations used for systemic effects. 1992. Available from: http://faculty.ksu.edu.sa/64448/Documents/Guideline%20BA%20-%20BE%20Part%20A.pdf.
9. Canada, Health Canada, Notice to Industry Removal of Requirement for 15% Random Replicate Samples, 24 Sept 2003.
10. Arnold ME. Implications of differences in bioanalytical regulations between Canada, USA and South America. Bioanalysis. 2011;3(3):253–8. doi:10.4155/bio.10.187.
11. USA, Electronic Records; Electronic Signatures. 1978. Code of Federal Regulations Part 11.
12. USA Food and Drug Administration. Draft guidance for industry: bioanalytical method validation. 1998.
13. Shah VP, Midha KK, Findlay JW, Hill HM, Hulse JD, McGilveray IJ, et al. Bioanalytical method validation—a revisit with a decade of progress. Pharm Res. 2000;17(12):1551–7.
14. USA Food and Drug Administration. Guidance for industry: bioanalytical method validation. MD, USA: Rockville; 2001.
15. Brazil ANVISA. Resolution RDC 27, minimum requirements for bioanalytical method validation used in studies with the purpose of registration and post-registration of medicines. Brazil; 2012.
16. Viswanathan CT, Bansal S, Booth B, DeStefano AJ, Rose MJ, Sailstad J, et al. Workshop/conference report—quantitative bioanalytical methods validation and implementation: best practices for chromatographic and ligand binding assays. The AAPS J. 2007;9(1): E30–42. doi:10.1208/aapsj0901004.
17. Fast DM, Kelley M, Viswanathan CT, O'Shaughnessy J, King SP, Chaudhary A, et al. Workshop report and follow-up—AAPS workshop on current topics in GLP bioanalysis: assay reproducibility for incurred samples—implications of crystal city recommendations. Aaps J. 2009;11(2):238–41. doi:10.1208/s12248-009-9100-9.
18. European Medicines Agency. Guideline on Bioanalytical Method Validation. 2011. Available from: http://www.ema.europa.eu/docs/en_GB/document_library/Scientific_guideline/2011/08/WC500109686.pdf.
19. European Medicines Agency. Reflection paper for laboratories that perform the analysis or evaluation of clinical trial samples. 2012. EMA/INS/GCP/532137/2010.
20. Canada, Conduct and Analysis of Comparative Bioavailability Studies. (12-105972-31), 22 May 2012. Available from: http://www.hc-sc.gc.ca/dhp-mps/alt_formats/pdf/prodpharma/applic-demande/guide-ld/bio/gd_cbs_ebc_ld-eng.pdf.

21. Japan, MHLW, Guideline on bioanalytical method validation in pharmaceutical development. 2013.
22. Japan, MHLW, Guideline on bioanalytical method (ligand binding assay) validation in pharmaceutical development, 2014.
23. USA Food and Drug Administration, Draft guidance for industry: bioanalytical method validation. 2013.
24. Global Bioanalytical Consortium. [cited 2016 April 7]. Available from: http://www.globalbioanalysisconsortium.org/.
25. AAPS Journal themed issue: best Practices for bioanalytical methods: recommendations from the global bioanalysis consortium. Available from: http://pharmagateway.net/ThemedIssuePage.aspx?JournalID=12248&CategoryID=1312.
26. USA Food and Drug Administration. Guidance for industry: safety testing of drug metabolites. 2008.
27. ICH Topic M 3 (R2). Non-clinical safety studies for the conduct of human clinical trials and marketing authorization for pharmaceuticals. 2009.
28. Health Canada Notice: clarification of bioanalytical method validation procedures, 8 Oct 2015, File number: 15-110985-741.
29. Mire-Sluis AR, Barrett YC, Devanarayan V, Koren E, Liu H, Maia M, et al. Recommendations for the design and optimization of immunoassays used in the detection of host antibodies against biotechnology products. J Immunol Methods. 2004;289(1–2):1–16.
30. Gupta S, Indelicato SR, Jethwa V, Kawabata T, Kelley M, Mire-Sluis AR, et al. Recommendations for the design, optimization, and qualification of cell-based assays used for the detection of neutralizing antibody responses elicited to biological therapeutics. J Immunol Methods. 2007;321(1–2):1–18.
31. Guidance for Industry, Immunogenicity Assessment for Therapeutic Protein Products. U.S. Department of Health and Human Services. Food and Drug Administration. Center for Drug Evaluation and Research (CDER). Center for Biologics Evaluation and Research (CBER). August 2014.
32. Gupta S, Devanarayan V, Finco D, Gunn GR 3rd, Kirshner S, Richards S, et al. Recommendations for the validation of cell-based assays used for the detection of neutralizing antibody immune responses elicited against biological therapeutics. J Pharm Biomed Anal. 2011;55(5):878–88. doi:10.1016/j.jpba.2011.03.038.
33. Committee for Medicinal Products For Human Use (CHMP). Guideline on immunogenicity assessment of biotechnology-derived therapeutic proteins. April 2008.
34. Guidance for the Industry. Assay development for Immunogenicity Testing of Therapeutic Proteins. Draft Guidance. U.S. Department of Health and Human Services. Food and Drug Administration. Center for Drug Evaluation and Research (CDER). Center for Biologics Evaluation and Research (CBER). December 2009.
35. European Medicines Agency. Guideline on immunogenicity assessment of biotechnology-derived therapeutic proteins. Draft. 2016.
36. Shankar G, Devanarayan V, Amaravadi L, Barrett YC, Bowsher R, Finco-Kent D, et al. Recommendations for the validation of immunoassays used for detection of host antibodies against biotechnology products. J Pharm Biomed Anal. 2008;48(5):1267–81. doi:10.1016/j.jpba.2008.09.020.
37. Mikulskis A, Yeung D, Subramanyam M, Amaravadi L. Solution ELISA as a platform of choice for development of robust, drug tolerant immunogenicity assays in support of drug development. J Immunol Methods. 2011;365(1–2):38–49. doi:10.1016/j.jim.2010.11.011.
38. Butikofer L, Lemaillet G, Faust H. Strategies to estimate and improve drug tolerance in anti-drug antibody assays. Bioanalysis. 2012;4(16):1999–2012. doi:10.4155/bio.12.174.
39. USA, FDA, Clinical Laboratory Improvement Amendments (CLIA). Available from: http://www.fda.gov/MedicalDevices/DeviceRegulationandGuidance/IVDRegulatoryAssistance/ucm124105.htm.
40. USA, FDA, Draft Guidance, Framework for Regulatory Oversight of Laboratory Developed Tests (LDTs) (2014).

41. USA, FDA, Premarket Notification 510(k). Available from: http://www.fda.gov/medicaldevices/deviceregulationandguidance/howtomarketyourdevice/premarketsubmissions/premarketnotification510k/default.htm.
42. Verhaeghe T, Barton HH, Hara H, Hucker R, Kelley M, Picard F, et al. Recommendations from the Global Bioanalysis Consortium Team A8: documentation. Aaps J. 2014;16(2): 240–5. doi:10.1208/s12248-013-9556-5.
43. Booth B, Arnold ME, DeSilva B, Amaravadi L, Dudal S, Fluhler E, et al. Workshop report: Crystal City V—quantitative bioanalytical method validation and implementation: the 2013 revised FDA guidance. Aaps J. 2015;17(2):277–88. doi:10.1208/s12248-014-9696-2.

Chapter 3
Logistical and Operational Practice in the Regulated Bioanalysis Laboratory

Patrick Bennett and Richard M. LeLacheur

Abstract This chapter provides information on the internal and external logistics and practices required to operate a regulated bioanalytical laboratory. Despite commonality afforded by health authority guidances, the organizational structure of laboratories conducting regulated bioanalysis varies across the industry. For example, an internal bioanalytical laboratory operating within a pharmaceutical company is likely to be structured and operate significantly differently from a contract research organization (CRO) laboratory focused on the same discipline. From personal experiences, we can attest to the potential benefits for different operational structures, tools, and practices depending on the size and geographical footprint of a bioanalytical organization. Particularly, as a bioanalytical laboratory grows, the need to adapt to the scale of data handling, information management, and associated communications requires operational structures to evolve accordingly. These and other variables discussed in this chapter demonstrate the operational and logistical differences between different types of bioanalytical laboratories. Despite the structural differences, there are also some common logistical and operational requirements for all regulated bioanalytical laboratories including: (1) Information Technology (IT) systems that provide security, data management, and automation, (2) Standard Operating Procedures (SOPs) and policies that drive both regulated and business activities, (3) metric tracking that assists in both business operations and scientific operations, and (4) document and sample lifecycle management.

Keywords Laboratory organization · Standard operating procedures (SOP) · Laboratory information technology (IT) · Metrics

P. Bennett (✉)
PPD Laboratories, 2244 Dabney Road, Richmond, VA, USA
e-mail: patrick.bennett@ppdi.com

R.M. LeLacheur
Agilux Laboratories, Inc., One Innovation Drive, Worcester, MA 01605, USA
e-mail: RLeLacheur@agiluxlabs.com

M.L. Rocci Jr. and S. Lowes (eds.), *Regulated Bioanalysis: Fundamentals and Practice*, AAPS Advances in the Pharmaceutical Sciences Series 26,
DOI 10.1007/978-3-319-54802-9_3

Abbreviations

ADC	Antibody-Drug Conjugate
BMV	Bioanalytical Method Validation
CAP	College of American Pathologists
CAPA	Corrective and Preventive Action
CLIA	Contract Laboratory Improvement Amendments
cGMP	Current Good Manufacturing Practices
COA	Certificate of Analysis
COTS	Commercial Off-The-Shelf
CRO	Contract Research Organization
ELN	Electronic Laboratory Notebook
EMA	European Medicines Agency
GCP	Good Clinical Practice
GCLP	Good Clinical Laboratory Practice
GLP	Good Laboratory Practice
HIPAA	Health Insurance Portability and Accountability Act
HPLC	High-Performance Liquid Chromatography
HRAM	High-Resolution/Accurate Mass spectrometry
IND	Investigational New Drug Application
IT	Information Technology
LIMS	Laboratory Information Management System
KPI	Key Performance Indicator
LC–MS	Liquid Chromatography–Mass Spectrometry
MD	Method Development
PK	Pharmacokinetic
QA, QAU	Quality Assurance, QA Unit
QC	Quality Control
SOP(s)	Standard Operating Procedure(s)
TAT	Turnaround Time
UHPLC	Ultra high-Performance (or Pressure) Liquid Chromatography
US FDA	Food and Drug Administration of the United States of America

3.1 Introduction

This chapter provides information on the internal and external logistics and practices required to operate a regulated bioanalytical laboratory. Because the bioanalytical industry is quite diverse, the chapter describes the industry structure as an introduction to how the laboratories may be organized. Despite commonality afforded by health authority guidances, the organizational structure of laboratories conducting regulated bioanalysis varies across the industry. For example, an internal bioanalytical laboratory operating in a pharmaceutical company is likely to be structured and operate significantly differently from a contract research organization (CRO) laboratory focused on the same discipline. From personal experiences we can attest to the potential benefits for different operational structures, tools, and practices depending on the size and geographical footprint of bioanalytical organizations. Particularly as a bioanalytical laboratory grows, the need to adapt to the scale of data handling, information management, and associated communications requires operational structures that evolve accordingly. These and other variables discussed in this chapter demonstrate the differences between bioanalytical laboratories in terms of operations and logistics. Despite the structural differences there are some common logistical and operational requirements for all regulated bioanalytical laboratories including (1) Information Technology (IT) systems that provide security, data management, and automation; (2) Standard Operating Procedures (SOPs) and policies that drive both regulated and business activities; (3) metric tracking that assists in both business operations and scientific operations; and (4) document and sample lifecycle management.

In addition to the regulated bioanalysis operational considerations above, a Quality Assurance (QA) unit that is independent of the laboratory is also required. QA audit and review are essential to ensure compliance with internal requirements such as those defined in SOPs and with relevant external requirements such as those of regulatory agencies. While a detailed description of the QAU is beyond the scope of this chapter, a constructive relationship between laboratory and QA personnel in which each group is trusted and respected by the other, is essential to a sustainable laboratory.

3.2 Pharmaceutical Bioanalytical Industry Structure

3.2.1 CRO Versus Pharma

There are two primary organizations in the pharmaceutical bioanalytical industry. The first are the pharmaceutical and biopharmaceutical companies. The second are CROs, which provide a variety of services across all phases of drug research and development. CROs can be limited to providing only bioanalytical services, or may also provide clinical, preclinical, pharmacokinetic, drug discovery, consulting, and/or other laboratory services.

Both pharmaceutical companies and CROs divide their bioanalytical operations into various segments. The choice of structure is usually a function of the company or department's size, mission, mandate, or management preference. These are discussed in the next sections.

3.2.2 Regulated Versus Non-regulated

Bioanalytical groups often differ by whether they perform regulated versus non-regulated workflows. Pharmaceutical companies typically have discrete bioanalytical functions within discovery, preclinical, and clinical groups that correlate with the extent of regulatory compliance needed. The discovery-based bioanalysis that is not governed by regulatory requirements is usually performed by different staff than preclinical and clinical bioanalysis where good laboratory practice (GLP), good clinical practice (GCP), good clinical laboratory practice (GCLP), and bioanalytical method validation (BMV) guidance characterize the work performed. Discovery-based bioanalytical projects tend to be more focused on obtaining an adequate result or answer in the least amount of time. Preclinical and clinical bioanalytical projects are usually performed in a much more structured manner with a major focus on regulatory compliance, consistency, method robustness, and confidence in the resulting data.

3.2.3 Discovery

Discovery bioanalysis groups prefer to develop more generic methods that can be used for an entire class of compounds. The amount of customization a discovery method may require depends on the analyte behavior, the number and types of analytes, the matrices collected, when the data is needed and what the data is to be used for. This approach falls within the "fit for purpose" paradigm in which the quality of the method should match the intended use of the data. More important projects—e.g., final pharmacokinetic (PK) studies to nominate a drug candidate for IND—will require more effort on method development and the extent to which the method is qualified. A method qualification is performed to characterize how well a method will quantify a given analyte and usually focuses on accuracy, precision, linearity, linear range, analyte carry-over and some limited stability experiments. Less important assessments—e.g., screening experiments—have reduced effort to customize a method and typically use methodology that is generic to a class of compounds.

3.2.4 Preclinical and Clinical

Bioanalytical functions supporting preclinical and clinical projects are producing data that may be submitted to regulatory agencies as part of drug filing applications. As such, the work falls into the category of regulated bioanalysis. The workflows performed include method development, method validation, and sample analysis. CROs may divide these functions into groups that are specific to one or two of these workflows. Method development and method validations are often performed in a dedicated group with very experienced scientists. These methods are then transferred to a group of scientists that are experienced with study sample analysis. Other CROs and most pharmaceutical companies assign method development, method validation, and sample analysis for a given project to a single individual and/or team, thereby ensuring that knowledge gained in method development and validation is retained and applied through sample analysis.

The bioanalytical methods developed for preclinical and clinical sample analysis require significant and formal method validation. The types of experiments performed and their acceptance criteria are described in regulatory agency guidelines, white papers, and publications. These have been covered in more detail in Chap. 2. Bioanalytical laboratories in both pharmaceutical companies and CROs that focus on preclinical or toxicology analysis may also have different groups performing preclinical and clinical analysis. While it seems that a validated method for preclinical species might be easily applied to validate a method for human samples, this is frequently not the case. Human control matrices as well as those from volunteers/patients are quite heterogeneous compared to those from animals. This is because age, healthy versus diseased, stage of disease, diet, etc. can impact the precision and accuracy of a method. Other challenges or differences include the volume of sample available, the metabolites observed, rate of metabolism, and observed concentrations of drugs or biomarkers.

3.2.5 Biopharmaceuticals

Increasingly, there may be groups dedicated to the bioanalysis of biopharmaceutical drugs (biologics). These include peptides, proteins, antibodies, oligonucleotides, and antibody drug conjugates. These drugs present different bioanalytical challenges relative to small molecule drugs. The differences between the bioanalysis of biologics relative to small molecules include how the method is developed, the reagents, equipment, instrumentation, software, expertise, sample handling, reference standards, and SOPs.

The reference standard material used to prepare calibrators and spiked quality control (QC) samples is very important to bioanalysis. In the case of small molecule bioanalysis, the purity and potency testing information contained on a certificate of analysis (COA) for the reference material allows definitive quantitation. This reference material is very homogeneous with very little difference between lots. In contrast, biopharmaceuticals are heterogeneous and can have measurable differences between lots. Therefore, an increased interaction between the formulation and cGMP groups that perform lot release with bioanalytical laboratories is required to ensure the appropriate lot of reference material, applied concentration value, and primary components are used.

3.2.6 Biomarkers

Paralleling the growth in biopharmaceutical bioanalysis is the formation of groups dedicated to the measurement of molecular biomarkers. While College of American Pathologists (CAP), Clinical Laboratory Improvement Amendments (CLIA), and exploratory biomarker methods have been routinely used, health authority guidance language is also driving the development of some biomarker assays within bioanalytical laboratories. These methods can be regulated or non-regulated and subsequently require various amounts of method validation or qualification. In some cases, this testing can be more extensive than regulated method validations for xenobiotic drugs, especially in cases with sample stability issues, matrix effects related to using surrogate matrices, and concentrations of endogenous analyte. Because biomarkers may be included in all phases of drug discovery, research, and development, bioanalytical scientists may have to interact with discovery groups, clinical operations groups, translational medicine groups, and central labs that influence modern drug development.

3.3 Information Technology Systems (Including Laboratory Information Management Systems)

Information technology (IT)-based systems are broadly implemented and critical to a regulated bioanalytical laboratory. They are composed of Laboratory Information Management Systems (LIMS), Electronic Laboratory Notebooks (ELN), enterprise business systems, communication, security, document management, and record keeping. While many pharmaceutical companies have business enterprise software, it is not widely used within their scientific research departments. However, within CROs, business-based software is commonly used. This includes sales management systems (e.g., Salesforce.com software), enterprise resource planning systems, and custom-built business management tools.

3.3.1 Laboratory Information Management Systems

LIMS systems can be commercial off-the-shelf (COTS), in-house built/proprietary, or a combination of the two. There may be multiple LIMS used within an organization depending on the type of research being performed, the types of instrumentation and data obtained, the type and complexity of data processing, as well as the historical progression of the laboratory. Different LIMS are typically used for different departments or divisions. This is true for both pharmaceutical companies and CROs.

The primary use of a LIMS system in bioanalysis is to manage study information and generated data. This includes the following:

- creation of the study design
  - number of animals, patients or volunteers, dosing schedules, and the dates and times of sample collection

- study sample chain of custody, use, and storage
- reference material chain of custody, use, and storage
- documenting and storing study sample analytical and related data
  - sample analysis batch creation
  - assay details (e.g., number of replicates and concentrations of standards and QCs)
  - run acceptance or failure
  - overall assay performance within and between studies
- quantitation calculations from imported raw data (e.g. data regression)
- data management
  - study sample results
  - what analytes have valid results
  - what samples need to be repeated with a different dilution factor
  - incurred sample repeats
  - selection of what result is reported if a sample was reassayed and has multiple valid results
- reporting and archival.

3.3.2 *Electronic Notebooks*

There are a variety of electronic notebook packages offering various and sometimes different capabilities or focus. Some are more process driven and geared for repetitive and highly regulated tasks such as those in cGMP laboratories for lot release. Others offer extreme flexibility and are less process driven. The key benefits of ELNs are their scalability and that they can bring all practitioners to a consistent approach, thus supporting compliance in a regulated laboratory. From an organizational standpoint, one or more ELN can be implemented for specific targets, departments, or enterprise-wide operations. Bioanalytical laboratories need a balance of fixed, process-based functions and operational flexibility in an ELN system. These needs may require more than one ELN or identification of an ELN that meets the majority of needs.

The process-oriented capabilities allow the lab to “lock down” routine processes, documents, forms, and procedures. Reagent preparation, for example, requires documentation of the materials used, the preparation procedure, lot numbers and expiration dates, storage conditions, volumes, containers, scientist, dates, and locations. Using a lab notebook could result in missing or incorrect documentation and an unlimited way of describing the preparation. A process-driven approach requires users to follow each step of the process and enter related details exactly the same way each time. In addition, the bioanalytical lab is tasked with accommodating new methodology with each new assay. Therefore, a bioanalytical ELN tool needs to have the associated flexibility to enable accurate recording of what are effectively routinely changing processes.

If sufficient flexible capabilities are available, the same ELN can be used for method development and similar research-based projects. This allows electronic documentation and traceability of bioanalytical methods, method development experiments, validation experiments, and other activities that change daily or from project to project.

3.3.3 Electronic Records and Signatures

Electronic signature is widely used in the financial industry to ensure privacy, security, and traceability, and it has been adopted by the pharmaceutical industry for similar reasons. In its basic form, an electronic signature is analogous to a "wet ink" written signature. Electronic signatures can be an electronic replica of a written signature or a purely digital record that is unique and traceable to a named individual. Either way, electronic signatures require a password to apply to a document. Electronic signatures are a requirement for use of electronic data and electronic submissions to regulatory agencies.

The two main types of data and/or documents in bioanalytical laboratories are electronic and paper. Both of these media must be managed in a careful and controlled manner. All laboratories are, or should be, continuously migrating toward a higher level of electronic-based records. When paper-based records are required, the ability to convert them to electronic media (e.g., pdf or similar) is highly desirable. Electronic records are an essential reality of bioanalytical laboratories.

In the U.S., electronic records fall under the 21 CFR Part 11 guidance. While not all electronic records require electronic signatures, great care and thought are needed when considering what electronic records are maintained and how they are handled within a laboratory operation. This process is part of the data life cycle. However, not all electronic records are study data. Both pharmaceutical company and CRO bioanalytical labs have many business functions that create electronic records that must be considered.

3.3.4 Security

There are two types of computer systems used in bioanalytical laboratories—open systems and closed systems. Open systems have operations that are portable, widely accessible from various computers and locations, and interact with other systems both internal and external to the laboratory. Closed systems are typically dedicated to a specific function or instrumentation. In bioanalytical laboratories, the latter is usually the system that controls the analytical instrument and collects the raw data from the detector. Appropriate selection of security practices and SOPs should be based on whether a system is open or closed and the risks associated with each.

Both scientific and business data should be considered at risk. Ensuring the security of this data should be a high priority in any organization. A pharmaceutical company's scientific data, drug portfolio, and research efforts must be secured from competitors or other malicious actions. In addition to the above precautions, CROs must protect their customer's information. This includes masking the customer's company name; the drug codes/names, study/protocol identities, etc. Furthermore, HIPAA-related patient confidentiality must be maintained according to individual region or country laws.

3.3.5 *Data Life cycle*

The life cycle of the various data types created by a bioanalytical lab must be mapped and procedures developed to allow continuity. The life cycle for electronic versus paper is similar with the exception that multiple copies of paper are not often created while multiple versions of electronic data are often created and appropriately tracked and differentiated.

Typically, data is obtained or created at one "location" and then consolidated with other data related to a specific study, project, or task type. Paper data should follow SOPs for how they are created, controlled, tracked, approved if necessary, and placed in either short- or long-term archival. Typically, long-term archival is a remote and professionally managed archival facility. Electronic data is typically backed up either immediately or following a specific schedule to a secondary server or similar centralized location. The data is then consistently backed up with redundancy up to the time that it needs to be archived. For both paper and electronic data, active data is typically stored in an easily accessible manner—i.e., secured paper filing system or online for electronic data. Once this period of time has passed per an SOP, it should automatically or manually be transferred to an off-line/off-site storage facility.

Most data storage includes three storage points:

(1) Immediate use. In this situation, the data is needed for further processing, review, or inclusion in a report.
(2) Long term. Long-term storage is usually easily available by all users that may need access to the data. It is secured, traceable, and retrievable.
(3) Archival. Archived data is usually removed from any "on-line" functions. In order to access the data, a formal request must be made to IT or similar party to restore the data onto a live/active system so that it can be reviewed or reprocessed as needed.

3.3.6 Other Software

The simplest way to view software in a bioanalytical laboratory is regulated or non-regulated. Regulated software includes any system that acquires, processes, stores, or manages data. Non-regulated software includes business software such as most word processing, presentation, accounting, and sales packages. Spreadsheets are unique in that they can be used in both modes. Care should be taken when using spreadsheets in a manner that is, or could be, considered as creating raw data, processing data, or similar.

Either departmental and/or company SOPs should describe the procedure for identifying how systems are used (i.e., regulatory compliant, business compliant, or neither), if they require validation, as well as the extent of and requirements for validation.

3.3.7 System Validations

Computer systems and the processes surrounding them must be validated to ensure they meet the intended purpose. This is a formal process that follows departmental and/or company SOPs. The intent is to test the system for all "mission critical" aspects—e.g., security, data collection, system stability, capacity, capability, communication with other systems, and user interfaces.

System validation SOPs should identify what systems should be validated, as well as the process and responsibilities associated with the systems. There are reasons to validate systems for both regulatory compliance as well as business requirements. The scope and approach may be different for each of these.

3.3.8 Information Technology SOPs

Standard operating procedures are required for proper use, maintenance, security, and lifecycle management of IT systems. Often there are multiple levels of SOPs in an organization. This includes corporate level, division level, and department level. Typically, corporate SOPs deal with global business and security processes. They are high level and guide all other SOPs. Divisional and department level SOPs typically address the requirements for the operations. These are more detailed and specific and may deal with regulatory compliance, business, finance, and science as well as the assignment of responsibilities. This is critical because of the multiple interfaces between a formal IT group and the wide variety of end users, software, hardware, locations, media, and laws and regulations.

3.3.9 System Lifecycle Management

IT systems should follow a system lifecycle SOP. This includes how systems are evaluated prior to purchase or creation and how they will be implemented, managed, maintained, and phased out. This is essential because it ensures the organization understands how various systems interact with each other and allows them to plan their resources appropriately. When a system is being phased out, the transitioning of the data from the legacy system to a new one is critical. Consideration of this concept even at the time of initial evaluation and purchase can save significant effort when the legacy system is deactivated.

3.3.10 Cross Organization (Within and Between Companies)

Most pharmaceutical and CRO bioanalytical laboratories do not operate independently. They are typically part of a larger organization, have partners, consultants, customers, preclinical and toxicology operations, clinical operations, biostatistics, or other CRO or third-party organizations. This requires the ability to share data between these groups in a controlled, traceable, and secure manner. Secure portals or encrypted email systems are typically used to meet these needs.

The information shared includes a variety of study data, patient/volunteer, or animal data. Clinical data including patient-specific information and analytical testing data must be merged. This data can come from a variety of locations and sources.

3.4 SOP and Document-Driven Operations

Formal controlling documents such as Standard Operating Procedures serve as a core element in the structure and operations of a regulated bioanalytical laboratory. These procedures serve to define requirements, conditions, and in some cases detailed processes across the entire operation, including, for example,

- Quality systems: creation, approval, revision of documents; management of deviations; training
- Information technology systems: network, workstation, major operating software; security; computer systems validations; lifecycle management including implementation of iterative software revisions and retirement of equipment

- Laboratory: key equipment, including calibration and maintenance specifications; requirements for analytical methods, including descriptive requirements of written methods and performance requirements of operating methods; documentation of reference material handling, use, and storage
- Data management: reporting of data for both method validation and sample analysis; repeat analysis data, including how to determine the final reportable result when multiple valid results exist.

When starting a new laboratory, the SOP infrastructure needs to translate the requirements for regulated bioanalysis as enumerated by the relevant governmental agencies into actions required internally. This infrastructure must be in place prior to the execution of any regulated analyses; the development of this infrastructure represents a major start-up expense for the new laboratory. In the long term, well-executed SOPs will provide return on that initial investment by helping to create a sustainable laboratory from both a data quality and operational perspective.

3.4.1 Laboratory Environment: CRO Versus Pharmaceutical Company

A bioanalytical laboratory is fundamentally a service organization regardless of where it is housed. Whether at a CRO or as a pharmaceutical company, regulated bioanalysis must be able to support filings with US FDA, European Medicines Agency (EMA), and other global regulatory agencies. CROs do benefit from the refined nature of their SOPs that result from input from numerous clients, their regulatory affairs consultants, and regulatory agencies. Consequently, CROs may have also very refined operations that are current with regulatory actions and trends. However, the operational processes of a lab within a CRO also must meet the needs of financial stability. Ultimately, the CRO must recognize that operational paralysis will occur if it attempts to satisfy all nuances, perspectives, and recommendations, and thus limits must be set on what is captured by the SOPs.

Conversely, a laboratory housed within a pharmaceutical company effectively serves a single client, and the SOP infrastructure will likely not be reviewed as frequently or by as many different organizations or individuals. These SOPs can more clearly reflect the initial requirements of a small number of internal authors. External review by one or more knowledgeable consultants can be invaluable in providing critical review and perspective, and effectively improve the SOPs. In time, the modifications proposed by regulatory agencies will need to be captured as well.

Ultimately, for both CRO and internal pharmaceutical laboratories performing regulated bioanalysis, the final clients are the governmental regulatory agencies.

SOPs must meet the needs defined in their guidance documents, and the input from these agencies that results from audit of the SOPs and resulting data needs to be taken seriously by the laboratory.

3.4.2 *Guiding Documents: SOPs Versus Policies and Guidelines*

Whether a laboratory is an independent company or embedded in a larger organization, guiding documents will be required to address a broad range of topics, such as

- Training: content and documentation of training
- Equipment: qualification, validation, calibration, maintenance requirements
- Analytical methods: validation, sample analysis, data management specifications, possibly encompassing both regulated and, separately, non-regulated bioanalysis
- Financial practices: accounting requirements for capturing expenditures and revenues (where applicable)
- Human resources requirements: confidentiality, non-discrimination, benefits
- Safety
- Communication/Reporting
- Security.

Multiple classes of documents may be created to accomplish this task. The role of each document class should be clearly defined, and care must be exercised to avoid conflict between documents. The following is an example of possible classes of documents and their scope:

3.4.2.1 Standard Operating Procedures

Relatively narrowly defined documents address a specific topic or task. An SOP should include sufficient detail to permit a trained user to execute the task in a compliant manner. The focus of SOPs should be on meeting the requirements to support data that will be submitted to regulatory agencies.

3.4.2.2 Policies

Policy documents may be broad, and define general requirements of the laboratory and organization. For example, policy documents may define safety expectations and practices within the organization, and a laboratory SOP may define steps required for safe operations of a specific piece of equipment.

3.4.2.3 Guidelines

A laboratory can perform both regulated and non-regulated bioanalysis. While the former is clearly controlled by SOPs, the latter may still benefit from clear requirements to help ensure uniformity, but also greater flexibility to address varied circumstances and objectives. Guideline documents can address the needs of non-regulated bioanalysis by defining, for example, assay performance requirements in early compound screening studies differently from requirements in final formulations and dose range-finding studies. All of these data are designed to support internal decision-making, and not regulatory filings.

3.4.3 Document Control

For each of the document categories defined, the requirements for document control must also be defined. For example, the initial creation of new documents may include

- Document numbering system and structure, often used to group-related documents and to include version control
- Document content, including required sections that may define scope, roles, definitions, etc.
- Document format, which may include the use of controlled templates to help manage defined content.

Beyond initial creation, all controlling documents need to be living entities, and defined procedures should exist that address requirements for,

- Periodic review and revision: Annual review of SOPs may be required, for example, compared with less frequent required review for Guidelines
- Document approval: defined roles for who effectively owns the document (author or responsible person), who reviews the document, and who approves the document
  - Defined process for changing the individuals who serve in those roles may be useful
- Issuance of documents: how electronic documents are controlled, secured, and accessed; how printed documents are managed
- Retirement of documents: how prior versions of existing documents will be secured and accessed. Work performed during the effective period of an SOP (prior to retirement) was defined and controlled by that revision, and should be reviewed and audited in that context.

As with most systems, foresight in defining these document control processes from the outset is beneficial to long-term management of the laboratory.

3.5 Metric Tracking

Key Performance Indicators (KPI, or simply metrics) are quantitative measures used to track operational performance of specific aspects of the laboratory or business. By placing a focus on a specific aspect of the operation, a metric effectively emphasizes important aspect(s) of the area being measured, which in turn may help drive improvement.

In defining what to quantify and track with a metric, thought should also be given as to what exactly needs to be tracked, and what price may be paid as staff work towards that specific objective. Consider, for example, a goal of fewer failed analytical batches. First, a definition of what "failed" means is required. Some possible outcomes for a failed analytical batch include

- Failure during extraction: splashing noted, 96-well plate dropped, etc.
- Failure during instrumental analysis: plate reader malfunctions and no reading are possible within the defined time limit, or a liquid chromatograph shuts down due to over-pressurization
  - For LC–MS analysis, failure may occur for the initial analysis, but the batch may be successfully analyzed via complete reinjection
- Failure occurs simply because the batch does not meet acceptance criteria: no calibration curve can be defined (too many standards fail to meet specifications), insufficient QC samples meet criteria, or blanks are not sufficiently clean.

For the purposes of a metric, which circumstances qualify as a failure? It is insufficient to track simple pass/fail, and instead, the metric should focus on tracking specifically what is valued. If laboratory staff is a major limiting resource, a pass/fail metric should focus on the productivity of their time and the ultimate pass or fail of their work. Conversely, if instrument time and productivity is tracked, a batch that needs to be analyzed twice uses significantly more time and reduces productivity. This batch is now tracked as two analyses, one fail and one pass. Note the difference in the statistics that would result from these two scenarios.

A second consideration is how the metric may drive undesirable actions and consequences. Consider the previously described situation again. If staff recognizes that the pass/fail metric is paramount in their laboratory, they may be motivated to ensure that batches pass—one way or another. The culture and values of the organization need to establish clearly the roles of metrics, and the scope of actions permitted in pursuit of metrics targets.

3.5.1 Objectives of Metrics

Metrics generally fall into two broad and complementary categories:

(a) Observational metrics are relatively high-level summaries of results or outcomes within the organization. While useful, a high-level metric may not indicate what change is needed if the performance level is unacceptable.

- Example: Batch pass/fail rate. While good to know, the metric does not help identify a predominant cause of failure that should be addressed. However, the long-term trend of pass/fail rate would generally be considered useful in monitoring the laboratory.

The objective of an observational metric may be to highlight a topic ("something's wrong here"), but not to help identify or drive specific corrective actions. Such metrics are useful in assessing the big picture and guiding the long-term performance of the laboratory.

(b) Actionable metrics are designed with specific tasks and improvements in mind.

- Example: System failure rate during instrumental analysis. Tracking the equipment failure rate for each of 10 LC–MS systems can help identify which system needs improved maintenance or repair (Fig. 3.1), especially when viewed over time (Fig. 3.2). MS-04 and MS-07 appear to have recurring issues that lead to significant downtime and make planning difficult.

The objective of an actionable metric is to indicate what specific corrective action would improve the outcome, or at least which specific area needs further investigation.

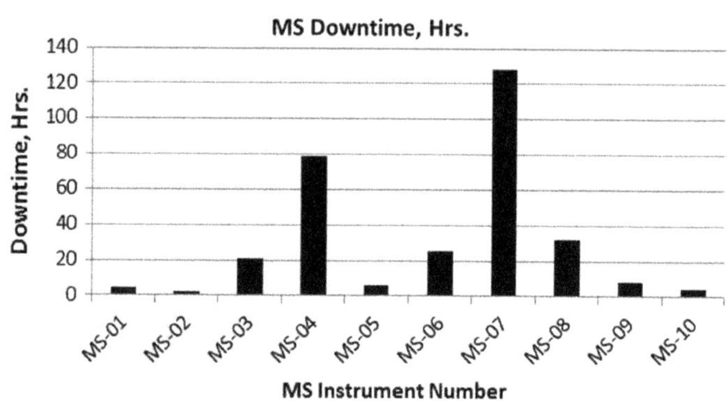

Fig. 3.1 MS downtime, Hrs

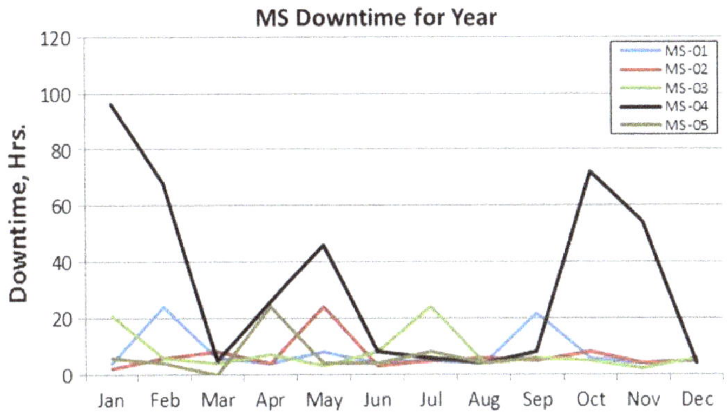

Fig. 3.2 MS downtime for year

3.5.2 *Defining Useful Metrics*

In simplest terms, a metric should only track what is important. For a new laboratory, a series of initial metrics may be defined and monitored. Over time, it will be noted that some metrics are frequently discussed, yet others are largely ignored. In this manner, the metrics help evaluate what is actually important in the organization.

Ultimately, useful metrics will often center on tracking core values and limiting resources. For example, a laboratory can value turnaround time, minimizing the time from receipt of a project to delivery of data. Another laboratory may value on-time delivery, in which a schedule is reliably met. Depending on circumstances and values, either can be worthwhile, and tracking against workload may be instructive (Fig. 3.3).

Tracking rate-limiting resources is a measure of efficiency, and can help guide actions such as purchasing or hiring. Due to the large expense of some modern bioanalysis instruments (e.g., LC–MS equipment, robotics), system time can be tracked as a limiting resource. Tracking utilization rates and outputs can help inform the laboratory when additional equipment should be purchased or how operational changes may maximize usage rates.

3.5.3 *Other Examples of Metrics*

Quality Metrics:

- QC/QA audit findings per study or task: Observation metric useful in discussions of long-term quality improvement.
- Audit findings by type or category (Fig. 3.4): Actionable metric useful to determine major problem areas. An example would be tracking which SOPs are

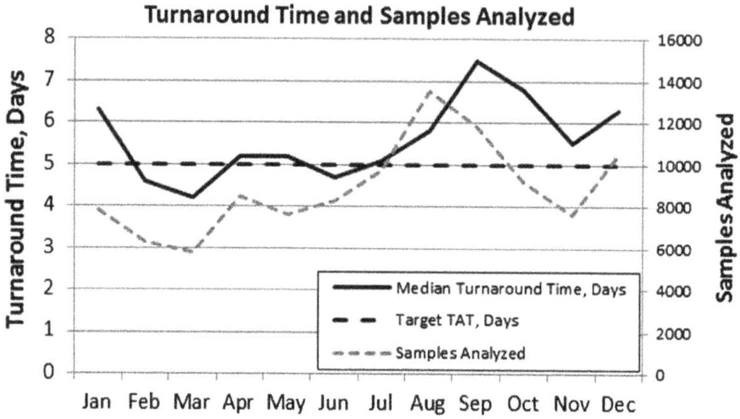

Fig. 3.3 Turnaround time and samples analyzed

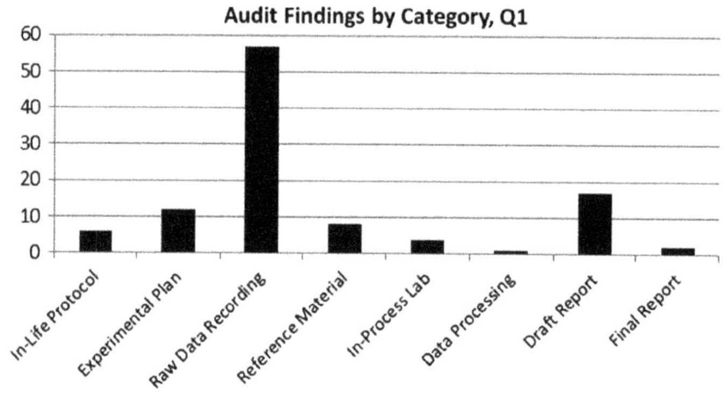

Fig. 3.4 Audit findings be category, Q1

violated most frequently. Additional training may be needed, an SOP may need to be revised to eliminate a troublesome (and not useful) requirement or a new process, automation, or personnel role may be required.

Financial Performance Metrics

- Revenue per employee or full-time equivalent, and revenue per instrument are common observational metrics in a CRO or any lab with financial tracking.

Efficiency Metrics

- Study sample results generated per instrument operation hour: observational metric that summarizes factors such as pass/fail rate, cycle time from injection to injection, batch content (fraction of batch that is study samples versus calibrations standards, QC samples, blanks).

- Revenue per instrument operating hour: a more detailed look at instrumentation that eliminates idle time (e.g., revenue per LC–MS or robotic system may be low due to a low workload, or due to inefficient utilization of the system).

Operations Metrics

- New method time: time from inception of method development (MD) to completion of validation and deployment of sample analysis method.
- Pass/fail rate for sample analysis: a counterpoint to previous metric. Excessive speed through MD may yield low-quality methods that fail frequently.
- Pass/Fail rate by cause: actionable metric that should highlight major causes of failure and drive remediation. Categories can include failure of instrument (hardware, software), analyst error, method deficiency, chromatography (peak shape, retention time), and QC performance.

3.5.4 Setting Target Performance Levels

Over the long-term, better performance is always desired. A more concrete goal such as hitting a 96% pass rate instead of simply asking for a higher pass rate may help drive performance improvement. Target levels are useful both in driving improvement and reaching performance that returns value on the investment.

For a new metric, an estimation of the appropriate target performance level should be considered carefully. An inappropriate target level can be counterproductive and demoralizing, and ultimately dismissed as unachievable. With time, tracking current performance may serve to inform everyone in the lab on the present status first, open a discussion on what is achievable and what is necessary, and help set a target that is both satisfactory and supported. Over the longer term, metrics and target performance levels should be re-evaluated periodically, revised where appropriate, and discarded where relevance is no longer apparent.

3.6 Organizational Evolution and Process Improvement

The bioanalysis of pharmaceuticals is constantly progressing. This progress is influenced by changes in the types of drugs being developed, the new or improved technologies, academic research, regulatory changes, and those driven by economics.

The same continuous process improvements used in manufacturing can apply directly to bioanalytical laboratories. The reasons behind the need for process improvement in bioanalytical laboratories also match the drivers for improvements in manufacturing—i.e., faster, better, cheaper, more innovation, increased competition, global markets, etc.

3.6.1 SOPs

SOPs are adjusted whenever there are changes in technology, business needs, and/or regulations. The implementation of new instrumentation, consumables/devices, or automation typically requires new or updated SOPs related to these changes. Similarly, regulatory changes will most likely also require new or updated SOPs depending on the extent of the changes.

SOPs can also be reviewed for process changes that can improve the laboratory's quality, productivity, or both. These can be driven by both business needs and the regulatory community.

Corrective and Preventative Action (CAPA) SOPs are an important part of process improvement. These formal programs assist in exploring the reason(s) for issues, establish corrective procedures, and assist in identifying issues and potential solutions proactively.

3.6.2 Analytical Approaches

The types of drug molecules being developed continuously evolve. Because of these changes, the technology used and the approach to analyzing the samples have also introduced changes. This includes the rapid growth of biologic drugs and the unique bioanalytical challenges they present. Bridging the boundary of small and large molecule drugs are the antibody drug conjugates (ADCs) and oligonucleotides that in themselves have fostered additional innovation on behalf of the bioanalyst. There has been a rapid growth of biologically derived drugs combined with smaller growth of small molecules as well as the increased complexity of molecules within the biologics and large molecule space. These include the types of antibody drug conjugates and oligonucleotides being developed.

Similarly, the increased use of molecular biomarkers to assist in the identification of important biological pathways, characterization of disease, potential drug effectiveness, ineffectiveness, safety, or toxicity is also changing the bioanalytical laboratory. Once the domain of the clinical chemistry diagnostic laboratory, endogenous compound biomarkers have now entered bioanalysis regulatory language. As such, the modern bioanalyst needs to be prepared to address these emerging needs with defendable science and application of the latest tools and techniques.

The above changes, requirements, and expectations have resulted in many technology shifts in bioanalysis. This includes the wide implementation of ultrahigh-pressure liquid chromatography (UHPLC) versus traditional HPLC systems, the continuous search for sensitivity and robustness from LC–MS systems, the growth of high resolution/accurate mass (HRAM) mass spectrometry, a rapid expansion of ligand-binding technologies, and the merging of ligand-binding capture with mass spectrometry detection.

Consumable products are also being improved or invented and introduced to assist in the changing requirements of bioanalytical sample analysis. These include the packing/coatings, dimensions, and performance of HPLC columns and sample extraction consumable devices. In addition, the increased use of products that can perform phospholipid removal; peptide, proteins, antibodies and oligo nucleotide chromatographic separations; immunological-based extractions and separations (i.e., bead based technologies); and enzyme technologies to assist in the digestion of biologics for convenient LC/MS analysis are indicative of the many advances now available to the bioanalyst.

3.6.3 Business

In parallel to the evolving science of drug research and development is the changing organization of the pharmaceutical industry. This includes the corporate structures, geographic locations of headquarters, operations and staff, CROs and other research partners, mergers, acquisitions, licensing agreements, and technology partnerships. The same economic pressures that apply to pharma companies also exist for CROs. As pharmaceutical companies increasingly partner with CROs, the structures of the CROs are also being forced to change in order to provide a broad service offering.

Both pharmaceutical companies and CROs (as well as academic and research institutes supporting pharmaceutical research) now operate within a regulated environment. These operations are rarely specific to any one country and are most often global. This requires organizations to learn how to operate under a variety of local and international regulatory guidelines, customers, projects, and their related challenges. Process improvement is not limited to the scientific challenges. The following areas are also important to include in process improvement: project management, sample logistics, all aspect of communication, data maintenance, and financial and legal aspects.

3.6.4 Automation

Automation is an integral part of process improvement. It can impact each part of an operation—IT, finance, compliance, lab operations, data management, communication, and sales management (for CROs). Automation's biggest impact is consistency, reduction of errors, reduced time to completion and, in many cases, a reduction in expenses. It does not always reduce the number of staff required, but does allow the shifting of staff to more challenging tasks and those that cannot be automated.

The most common automation in a bioanalytical laboratory is IT based and includes a LIMS system and, increasingly, ELN systems. These systems enforce consistency in how data is acquired, maintained, and used in a manner consistent with the operational SOPs. These software applications ensure standard approaches to sample identification, sample receipt, tracking, results, and reporting. Other IT-based software can monitor systems such as freezer and refrigerator storage conditions, archival rooms, or other facility temperature and humidity conditions, and access to secure locations. ELNs and related software can ensure processes are followed and documented accurately. This reduces the potential for performance errors as well as unnecessary interactions between groups—i.e., lab operations and QC, IT, QA, customers, and others.

Instrumentation or mechanical automation is also an important aspect to process improvement. Automating sample aliquoting, diluting, QC and calibration curve preparation, and other sample processing procedures can be implemented. These increase consistency, reduce staff engagement in routine tasks, improve documentation, and often increase throughput for a given task.

3.7 Sample Life Cycle

The types of matrices, collection methods, collection time points, collection devices, shipping, receiving, analysis, and archival logistics can be very complex. The life cycle of a sample is dependent upon the stage of development—e.g., discovery, preclinical, Phase I clinical, or later phase clinical—as well as its use—e.g., metabolism, pharmacokinetics, immunogenicity, and biomarkers.

In all cases a protocol is prepared that describes what, when, how, and how much sample is collected as well as the specific details of the sample collection process, collection device(s), sample processing, labeling/identification, short-term storage, intermediary locations, and final location for sample analysis. Following sample analysis, SOPs are typically followed for duration of storage and method of disposal and related documentation.

3.7.1 Types of Samples

Preclinical samples are used for discovery, toxicology, and various IND-related experiments. Typical sample types used for bioanalysis include brain, liver, and blood/serum/plasma. Depending on the purpose of the experiment, other sample types such as urine, feces, kidneys, and other organs may be collected. Very small volumes of blood/serum/plasma are typically collected and used specifically for bioanalysis.

Samples from clinical studies require special consideration. This includes informed consent documentation which dictates specifically what samples can (or can't) be tested for, as well as patient/volunteer confidentiality.

3.7.2 Collection

Samples are collected and their storage devices uniquely identified, processed as necessary to prepare plasma or serum, stored under appropriate and controlled temperatures, and shipped (or transferred) to a location for temporary storage or sample analysis. Detailed documentation of each step is typically performed.

3.7.3 Analysis

Study samples are formally received and logged into a LIMS system at the analytical facility. The samples are continuously maintained at an appropriate temperature in a controlled manner. Full chain of custody is usually maintained for regulated studies. This includes continuous temperature and security monitoring of the storage location. The time out of the formal storage location and related conditions are tracked—this includes conditions for thawing, time under-thawed conditions, storage location, and times samples were taken from and returned to the storage location, as well as the identification of the staff performing these tasks. Once analysis is completed and the results are reported, SOPs (or a contract if one exists) direct the duration and responsibilities for long-term storage, disposal timing and methodology, and the related documentation.

3.7.4 Geographic Considerations

Both preclinical and clinical samples can be collected anywhere in the world and analyzed in another location. This requires significant thought surrounding the validation of sample stability and shipping logistics. Samples that have stability issues require more attention and care in this process—as do those related to safety, rare diseases, time-constrained analytical testing (e.g., cellular or functional testing), and Phase I samples. There may also be customs or other governmental laws that dictate how samples are transferred from one country to another.

3.8 Conclusion

Maintaining a successful and vibrant bioanalytical laboratory requires an awareness of industry and technology changes, operational procedures, and metrics. This chapter provides an overview of the operational aspects of a regulated bioanalytical laboratory. This is often a mixture of innovation and consistency, but always under a banner of controlled processes. This is where the concepts of lifecycle management become important. These concepts along with well-written SOPs allow an organization to routinely assess the appropriateness of its current procedures and capabilities and make adjustments as the industry needs, technology availability, and regulatory requirements evolve. The ability to develop and use metrics to track capacity, operational excellence, issue tracking, and problem resolution create indicators of organizational performance. These metrics can assist in guiding an organization to processes that may need attention including adding capacity, increasing training, increasing automation, or adding new procedures.

References [1–4] serve as important and extensive reading about regulatory guidelines regarding selected aspects of regulated bioanalysis within the United States and Europe.

References

1. European Medicines Agency. Guideline on bioanalytical method validation 2011. http://www.ema.europa.eu/docs/en_GB/document_library/Scientific_guideline/2011/08/WC500109686.pdf. Accessed 4 April 2016.
2. U.S. Department of Health and Human Services, Food and Drug Administration. Guidance for industry: bioanalytical method validation May 2001. http://www.fda.gov/downloads/Drugs/Guidances/ucm070107.pdf. Accessed 4 April 2016.
3. 21 CFR Part 58—Good laboratory practice for nonclinical laboratory studies.
4. 21 CFR Part 11—Electronic records; electronic signatures—scope and application.

Chapter 4
Regulated Bioanalysis: Documentation and Reports

Surendra K. Bansal and Faye Vazvaei

Abstract The early Crystal City Bioanalytical workshops did not discuss the requirements for bioanalytical documentation and reports. The major bioanalytical guidance from FDA and EMA provided only broad outlines for the requirements in bioanalytical documentation and reports. The bioanalytical practitioners were therefore left to decide what to document in the reports and what to maintain in the archives for inspection. This has led to bioanalytical reports of various shapes and sizes. Parallel to the development of the bioanalytical guidance, ICH developed an electronic Common Technical Document (eCTD) guidance for preparing regulatory reports. Recently, FDA issued a mandatory guidance for submission of data and reports electronically for certain submissions. These latter changes have also been influencing how the bioanalytical data and reports would be submitted in the future. This chapter provides an overview of the bioanalytical documentation and reports for bioanalytical studies intended for regulatory submission.

Keywords Bioanalytical documentation · Data · Reports · Electronic submission · eCTD · SEND · SDTM

4.1 Documentation and Reports

Bioanalysis employed for the quantitative determination of drugs and metabolites in biological matrices provides the essential raw data for pharmacokinetic studies performed for research and development, and ultimately regulatory submissions for the approval of pharmaceutical products. Because the quality of pharmacokinetic data and reports is highly dependent on the quality of the underlying bioanalytical

S.K. Bansal · F. Vazvaei (✉)
Roche Pharma Research and Early Development, Pharmaceutical Sciences, Roche Innovation Center New York, 430 E 29th Street, New York, NY 10016, USA
e-mail: faye.vazvaei@roche.com

S.K. Bansal
e-mail: surendra.bansal@gmail.com

M.L. Rocci Jr. and S. Lowes (eds.), *Regulated Bioanalysis: Fundamentals and Practice*, AAPS Advances in the Pharmaceutical Sciences Series 26,
DOI 10.1007/978-3-319-54802-9_4

data, it is important that procedures and specifications for bioanalysis are standardized, especially for work performed to support the approval of pharmaceutical products. Standardization of bioanalytical procedures and specifications started in 1990 with a global bioanalytical conference, commonly known as the Crystal City Bioanalytical workshop/conference [1]. The subsequent series of Crystal City bioanalytical conferences were highly successful in standardizing the procedure and specifications for the process of bioanalytical method validations and sample analysis [2]. However, relatively little attention was paid to standardizing the documentation or report writing for bioanalytical work. There were no discussions on bioanalytical documentation or report writing at the first two Crystal City bioanalytical conferences [1, 3], prior to the release of the first comprehensive regulatory guidance on bioanalysis by the US Food and Drug Administration [4]. As a result, bioanalytical practitioners started to develop their own procedures for documentation and report writing. While the reports at that time generally provided an adequate description of the bioanalytical work and the results; the lack of standardization led to wide variations in bioanalytical documentation and report writing. This variation was noted by regulators and it made the regulatory reviews difficult, especially during the drug approval process [5].

In the regulated bioanalytical space, thorough documentation of bioanalytical activities is crucial because inadequate documentation and/or reporting of bioanalytical data may jeopardize or delay drug approval. Both Good Laboratory Practice (GLP) and Good Clinical Practice (GCP), under which regulated bioanalysis is performed, mandate contemporaneous documentation of laboratory activities and full reporting of the results [4, 6]. All study data (electronic and paper) generated to support drug application filings should be preserved, readily retrievable, and available for inspection by regulatory reviewers and inspectors. The documentation must be clear and complete, such that the study can be reconstructed with the archived information even in the absence of the personnel who performed the study.

Independent but parallel to the development of bioanalytical guidance, the International Council for Harmonisation of Technical Requirements for Pharmaceuticals for Human Use (ICH) has been providing guidance on harmonizing regulatory submissions through the Common Technical Document (CTD) and its electronic version. The instructions and requirements in CTD, common for all reports in regulatory submission, have also influenced the writing of bioanalytical reports. In this chapter we will cover both the documentation and reports for regulated bioanalysis.

4.1.1 Regulatory Guidance Perspective

The first FDA bioanalytical method validation (BMV) guidance [4] called for full documentation of bioanalytical work, and required that the data generated for bioanalytical method establishment (development and validation) and sample analysis should be available for data audit and inspection. The guidance also provided a description of what should be documented for bioanalytical method validation and sample analysis [4]. It categorized this documentation into four parts: (1) Summary information, (2) Method development and establishment, (3) Bioanalytical reports for samples analysis, and (4) Other information for bioanalysis.

The FDA guidance provided a broad outline of the required bioanalytical documentation and reports but did not specify what documentation should be provided in the bioanalytical reports and what should be maintained for inspection in the archives [4]. Therefore, the bioanalytical practitioners were required to make the decisions.

EMA's guideline on bioanalytical method validation [6], released in 2011 (and effective in 2012), also provided guidance on how to document bioanalytical work. The EMA guidance provided additional clarity on what to include in the method validation and bioanalytical reports. Table 4.1 provides a summary of this required information by each of these two major guidance documents. The guidelines provided by the two agencies for documentation and report writing are similar and complementary. Together, however, these guidance documents still fail to create a standard for outlining the specifics on what to include in bioanalytical validation or analysis reports, and what additional information should be archived for inspection/auditing purposes.

Subsequent to the FDA and EMA guidance, other regional regulatory agencies provided their bioanalytical documentation guidelines. The Japanese guidance [7, 8] provides an outline for documenting bioanalytical work, with one significant change from the requirements given in Table 4.1. This change specifies that only representative chromatograms are needed in the reports. The Chinese (CFDA) guidance [9] provides a brief description of what to document. CFDA requirements were similar to those provided in Table 4.1 with one difference—asking for 20% of sample chromatograms with no exceptions. The recent bioanalytical guidance included in the 2015 edition of the Chinese Pharmacopoeia also provides an outline of the required documentation and reports [10], but requires 100% of sample chromatograms to be included in the bioanalytical reports for bioequivalence-type studies. Bioanalytical guidances from ANVISA [11] and Canada [12] do not include specific information related to documentation requirements. The abbreviated bioanalytical guidance from most other countries provided as part of their bioequivalence/bioavailability study guidance do not give specific details on bioanalytical documentation or reports [2, 13]. Given the abbreviated nature of the guidance documents from these other countries, we advise that bioanalytical practitioners follow the requirements of the more comprehensive FDA and EMA guidance documents [4, 6].

The draft FDA bioanalytical guidance issued in 2013 [14] suggested only minor changes to the documentation requirements given in the current FDA guidance [4]. The draft guidance expanded the instructions on electronic format of the documented data, provided details on which tables to include in bioanalytical reports, and added some example data tables for electronic Common Technical Document (eCTD) filing. It also provided guidance on the documentation required for system suitability and system equilibration assessments.

Table 4.1 Specific guidance on documentation recommended for Validation and Bioanalytical (sample analysis) reports in FDA [4] and EMA [6] guidelines for bioanalysis

Common or summary information in Validation and Bioanalytical reports		
Topic	FDA 2001 guidance [4]	EMA 2011 guidance [6]
Source data	Good record keeping is an essential part of bioanalysis. Data should be documented and available for audit and inspection	All source data should be available in original format and available on request by regulatory authority
Deviation	Document any deviation from SOP, protocol, or GLP (if applicable) with reason and justification	Deviation from protocol, analytical procedure or SOPs should be recorded. Describe deviation, impact on study, and supportive data in reports
SOPs	General and specific SOPs are an essential part of bioanalysis	When detailed information is provided in the report a reference to analytical SOP is sufficient; otherwise append SOP(s) to the report
Reference standards	Provide source and lot number, expiration date, certificate of analyses when available, and/or evidence of identity and purity for reference standards	Provide origin, batch, certificate of analysis, stability, and storage conditions for reference standards
Summary information (Validation reports)	Summary table of validation reports (validation, partial revalidation, and cross-validations) in chronological order. Include assay method code, type of assay, and reason for new method or additional validation	Provide a summary of validation performance in validation reports
Summary information (Bioanalytical reports)	Summary table of assay methods used. The table should contain protocol number and title, assay type, assay identifier code, and bioanalytical report identifier code. Include a cross-reference table if multiple identifier codes are used for the same validation or bioanalytical report	No specific information for documentation provided*
Other information	• Lists of abbreviations, codes, and references • References and specific SOPs for performing bioanalysis	No specific information for documentation provided*
Validation reports		
Analytical method	An operational description of the analytical method	Details and source (literature reference if applicable) of analytical method. Details to include analyte, internal standard, sample pretreatment, extraction, and analysis

(continued)

Table 4.1 (continued)

Common or summary information in Validation and Bioanalytical reports		
Topic	FDA 2001 guidance [4]	EMA 2011 guidance [6]
Calibration standards and QC samples	No specific information for documentation provided*	Matrix and anticoagulant (if applicable), preparation details with dates and storage conditions
Analysis and results	• Description of experiments conducted to determine accuracy, precision, recovery, selectivity, LLOQ, calibration curve (equations and weighting function), and relevant data from these studies • Documentation of intra and inter-assay precision and accuracy • Legible annotated chromatograms or mass spectrograms if applicable	• Table of all analytical runs with analysis dates, whether run passed or failed and reason for failure • Table of calibration results of all accepted runs, including calibration range, response function, back calculated concentration, and accuracy • Table of QC results of all accepted runs (within and between run precision and accuracy). Mark values outside QC acceptance criteria • Data on selectivity, LLOQ, carryover, matrix effect if applicable, dilution integrity • Present all measurements with individual calculated concentrations in report • Provide run acceptance criteria, unexpected results obtained during validation with full justification of action taken
Stability	A description of stability studies and supporting data	Stability data of stock solution, working solution, and QC samples, covering applied storage conditions
Bioanalytical reports		
Assay procedure	Equations used for back calculation of results	Short description of assay procedure. Include a reference to applicable validation report(s)
Sample tracking	• Summary table on sample processing and storage, containing sample ID, collection dates, storage prior to shipment, information on shipment batch, and storage prior to analysis. Include dates, times, sample condition, and any deviation from protocols • Document missing samples	Dates of receipt and contents for sample shipments, sample condition on receipt, storage on location, and condition
Analytical runs	Table of analytical runs—include information on run ID, date and time of analysis, assay method, analysts, start and stop times, significant equipment and material changes and deviations or issues	• Table of all analytical runs and study samples with run dates and results • Identify failed analytical runs, assay dates and the reason for failure

(continued)

Table 4.1 (continued)

Common or summary information in Validation and Bioanalytical reports		
Topic	FDA 2001 guidance [4]	EMA 2011 guidance [6]
Calibration standards and QC samples	• Tables for calibration curve and QC data. Provide equation used for calibration • Summary information on intra- and inter-assay accuracy and precision for calibration curve and QC data used to accept analytical runs. QC graphs and trend analysis are encouraged in addition	• Table of calibration and QC results for all passing analytical runs • QCs outside acceptance criteria should be marked • Storage conditions of stocks and QC samples should be provided
Sample analysis results	Table of all study samples with identifying samples and analytical runs, raw data and back calculated results, integration, and other reporting codes	Table of all study samples with identifying analytical run, sample ID, run dates, and the concentrations obtained
Reassay	Reassay tables containing sample identifier, reason for reassay, original and reassay values, reported results, the requestor for repeat analysis, and manager authorizing reanalysis	Reassay tables containing sample identifier, reason for reassay, original, and reassay values
Reintegrated data	For reintegrated chromatographic data include initial and repeat integration results, methods for reintegration, reason, requestor and manager authorizing reintegration, and run ID	No specific information for documentation provided*
Incurred samples reanalysis	No specific information for documentation provided*	Results of incurred sample reanalysis may be supplied in validation, bioanalytical, or standalone report
Chromatograms	Complete chromatograms for 5–20% of subjects including calibration standards and QC samples. For bioequivalence studies provide chromatograms for 20% of serially selected subjects and 5% for other studies	For bioequivalence studies all chromatograms of runs for 20% of subjects including corresponding calibration standards and QC samples. Only representative chromatograms for other studies should be appended

*While no specific information for documentation was provided for this section in the listed reference, information for documentation should be derived from the required activities on the topic in the reference, and also from the draft revision of the guidance (14) and white papers

4.1.2 Crystal City Workshops Perspective

White papers from the Crystal City Bioanalytical conferences I and II [1, 3] were silent on bioanalytical documentation and reports. The white paper from the Crystal City III bioanalytical conference [15] was released after the 2001 FDA guidance on bioanalysis [4], and provided clarification on the documentation instructions given in the FDA bioanalytical guidance. This white paper contains a comprehensive table that gives detailed information on what should be presented in bioanalytical reports (method validation and samples analysis) and what supplemental information should be archived at the analytical site [15]. It further suggests that some of the supplementary information may be added to the report in appendices. For example, SOPs are considered part of onsite documentation, but could be optionally added in the report appendices [15]. The EMA guidance calls for the inclusion of references to the SOPs when information is given in the reports, but recommends adding the SOP in appendices when the information is not provided in the reports [6]. The FDA guidance does not provide any specific guidance on the inclusion of SOPs in bioanalytical reports [4].

The next opportunity for a Crystal City conference to weigh in on bioanalytical documentation came after the release of the FDA draft bioanalytical guidance in 2013. The Crystal City V white paper [16] summarizes discussions held at the workshop on the documentation sections in the FDA draft guidance [14] and the Crystal City III report [15]. It provides clarification on the data to be included in bioanalytical reports, and recommends that the format of data tables and reports should be easy for review. For electronic submissions, the white paper [16] asked for the inclusion of hyperlinks and bookmarks for easy reference. The white paper [16] provided clarity on the level of method development activities to include in the reports, and recommended that a synopsis of such activities in the validation report is sufficient. As methods are improved iteratively over time, and multiple validations performed, the synopsis in the validation report should capture the evolution of the method [16]. It was also suggested that the method history should be captured in the summary section of the eCTD to allow the reviewers to relate the data and methods from multiple studies in the filing [16].

4.2 Documentation Systems—Paper to Electronic

For documenting the activities performed during a study, bioanalytical laboratories use a variety of documentation systems, from paper-based to fully electronic, as given below.

- Bound laboratory notebooks: A typical bound and paginated laboratory notebook is a classical approach for documentation. It has the advantage of ensuring contemporaneous documentation of activities without a fear for deliberate or accidental erasure, removal, or ultimately loss of the recorded data. They can also be controlled centrally to allow additional management or QA oversight.

- Paper forms: Forms can be developed and preprinted with instructions on specific activities required according to the laboratory SOPs or the analytical method. The analyst can record directly on the form. Forms offer the analyst preprinted, stepwise instructions for performing the required activities, resulting in a higher level of compliance with the SOPs and the analytical method. However, because it is relatively easy to replace a form without leaving a trace for its removal and replacement, the regulated laboratory should ensure that the forms are highly controlled. The system should ensure that the forms cannot be replaced or destroyed. For the laboratories developing electronic notebooks (ELN), the preprinted forms are also useful templates for required activities.
- Hybrid paper and electronic documentation: In the hybrid paradigm, automated data and observations are captured electronically, and manual data is captured on paper. Manual activities are documented on paper forms or bound notebooks. If a paper-based system is primary, there should be a link or reference in the paper documents that leads to the location of recorded data on electronic media.
- ELN: Historically, bioanalytical activities have been recorded on paper. With the rise of technology and availability of electronic solutions for bioanalytical data capture, processing, reporting, and archiving, it is only a matter of time before laboratories migrate to electronic solutions for all activities and become fully paperless. Although progress is being made to create paperless laboratories, it is not yet the norm. When adopted properly, electronic notebooks can offer several advantages, including error alerts, efficient searching, archiving and rapid reviews.

4.3 Essential Documentation in Regulated Bioanalysis

GLP and GCP regulations require that all regulated studies are fully documented. For bioanalytical studies one must also comply with the guidelines in the regulatory bioanalytical guidance. This section covers the essential documentation required when performing regulated bioanalysis, and is not meant to list all documentation required for a GLP/GCP operation. The activities performed in bioanalysis must be fully and contemporaneously documented, such that the study can be reconstructed at a later time. The reconstruction of the study may be required after many years, for a regulatory review or inspection. All source data and documentation generated either electronically or on paper should be preserved. Major categories of bioanalytical activities requiring documentation and data collection are described below:

- SOPs
- Bioanalytical Study Plan
- Bioanalytical training
- Equipment and Instruments
- Bioanalytical Method
- Study Performance Activities.

4.3.1 SOPs

Laboratory management should ensure that SOPs are available to the staff for performing regulatory work. SOPs should be written by qualified individuals familiar with the bioanalytical procedures and regulatory requirements. The SOP should clearly describe the purpose, scope, responsibilities, and procedures to be followed within its scope. It should be approved by laboratory management and should be periodically reviewed throughout its life cycle. A review by the Quality Assurance Unit is also a good idea. A complete historical trace of the SOPs and their revisions (i.e., version control) must be kept, including the information on retirement of the SOP when obsolete.

In a global organization the question often arises: should the SOPs be global or local? The use of global SOPs ensures uniformity in operation for a global organization but local SOPs are easy to manage and administer. The local SOPs, written and administered by local management, tend to have a better compliance record and contribute to developing the local culture. Therefore, in order to leverage the advantages of each type of SOP, one should consider writing agreed upon global procedures into local SOPs, written and administered by the local management and scientific teams.

4.3.2 Bioanalytical Study Plan

A high level scope for sample analysis describing the laboratory, personnel, and workflow methodology can be specified in the nonclinical study plan or the clinical protocol. The study plan for nonclinical (GLP) studies can also contain instructions regarding sample collection, handling, and shipment. Specific and detailed instructions for sample handling in clinical (GCP) studies can also be described in a separate document, commonly known as the sample handling or laboratory manual.

A separate comprehensive bioanalytical plan can be written to describe the bioanalytical study and to identify the responsible personnel conducting the study. The comprehensive bioanalytical plan can include the details of the sample shipment(s), bioanalytical method, instructions for reporting data, and other relevant information for the study. Caution must be used when writing such independent plans for GLP studies. If such a bioanalytical plan is written, local or GLP regulation may require that it is forwarded to the Study Director to ensure one central control of the GLP study. The bioanalytical study plan has been referred to as 'work instruction' by the EMA guidance [6], and was recommended by the Global Bioanalysis Consortium (GBC) documentation team as optional for the work done at the in-house sponsor laboratory [17]. The GBC team recommended the use of a bioanalytical plan as a good practice for studies for which the bioanalytical phase is outsourced [17]. The comprehensive bioanalytical plan can also be written for a validation study. If a validation is conducted according to the GLPs, it is imperative that a validation plan is written according to the site's SOP on GLP implementation and that the validation study director is identified.

4.3.3 Bioanalytical Training

Management should ensure that the training of the personnel involved in performing regulated bioanalytical work is well documented. There should be an SOP governing the training and its documentation. Besides regulatory GLP/GCP and SOP training, procedural training to acquire a skillset and proficiency (e.g., operation of an instrument) must be conducted and documented according to the training SOP. Training on performing the assay should also be documented. An analyst is considered trained on the assay when they perform a full or partial validation, or by performing the analysis with the oversight of an analyst already trained on the analysis [14].

4.3.4 Equipment and Instruments

For regulated bioanalysis, it is essential that analytical instruments and computer systems utilized in performing the analysis generate reliable, consistent, and accurate data. Therefore, analytical instruments and computer systems including Laboratory Information Management System (LIMS) should be qualified or validated prior to use. U.S. Pharmacopeial Convention (USP) general chapter <1058> provides a comprehensive overview and an approach to qualify analytical instrumentation [18]. This USP chapter defines the instrument qualification process as consisting of four phases: Design Qualification (DQ), Installation Qualification (IQ), Operational Qualification (OQ), and Performance Qualification (PQ). The chapter further categorizes the analytical instruments into three groups according to their complexity. The phases and the level of effort made for qualification of an instrument depends on its group and complexity. As part of qualifying the instrument, standard operating procedures must be established and used in operation, maintenance, calibration, and change control. Certain instruments require routine maintenance and calibration as part of their PQ. All documents generated during the instrument qualification initially, and with ongoing performance qualification, should be available in the laboratory or archived.

Certain instruments require injection of samples to condition the instrument or to perform system suitability. The calibration standards or QC samples that are also part of the analytical runs should not be used for conditioning or system suitability, as they can be construed to bias the decision for acceptance or rejection of the run [16]. Use of clearly documented, separately extracted samples or extracts from completed runs is recommended for this purpose [16].

4.3.5 Bioanalytical Method

Bioanalytical methods are either developed in the laboratory, transferred from another laboratory, or obtained from the literature. The developed or transferred method should be written up according to the laboratory documentation system before starting the validation. Any modification to the written method should be documented. Major changes to the method may require authorization by the laboratory management as per applicable SOP, and may extend validation experiments until the method is considered fully validated. Once the method has been validated, it should be finalized as the version to be used for samples analysis.

After validation, the method can still undergo changes. These changes should be well documented in a version-controlled manner. The use of the specific versions should be documented within the respective studies for traceability. Significant post-validation changes to the method require partial validation, which should be reported as an amendment or addendum to the validation report.

4.3.6 Study Performance Activities

A typical regulatory bioanalytical study has a long list of activities that must be documented and archived at the end. The following list is meant to provide an example of the type of activities and study-related information generally documented during analysis of samples from a regulated study. Keeping such a list in a laboratory guidance document can be very helpful in ensuring regulatory documentation. The list is a reminder and can be adapted according to the assay types and laboratory systems. All activities and information must be documented contemporaneously in the laboratory notebook, whether paper or electronic.

Typical Study Performance Activities and Their Documentation

1. Key Study Identifiers
2. Standard Compounds
3. Materials and Supplies
4. Instruments
5. Assay Settings
6. Preparation of Reagents/Mobile phase
7. Preparation of Stock Solutions of Standard Compounds
8. Preparation of Calibration Standards and Quality Control Samples (QCs)
9. Sample processing
10. Pre-study Assay Evaluation
11. Sample Storage Information

12. Analytical Runs:
    - Pre-extraction documentation
    - During-run documentation
    - Post-run documentation.
13. Quality Assurance
14. Supplementary information and General Notes.

In documenting the above activities related to a study, one should keep in mind that the ultimate test for valid documentation is to be able to reconstruct the study in the absence of the personnel involved in performing the study. Typical documentation for each of the activity is described below.

1. **Key Study Identifiers**

Study number, study title, key personnel, and other important identifiers should be listed up front for identification and easy reference. Part of this information is also retained in the 'Master Schedule' generally maintained by QA for regulatory purposes.

- Study Number
- Study Title
- Analyte(s) Name
- Study Director (for nonclinical GLP studies)
- Bioanalytical Principal Investigator (PI or Lead Bioanalyst) with example signatures and initials
- A listing of all personnel contributing to the bioanalytical study with example signatures and initials
- Analytical Method Number or other reference to a detailed documented analytical procedure.

2. **Standard Compounds**

Bioanalysis is a comparative science; the data obtained relies heavily on the quality of the compound(s) used as reference standard. The following information for reference standards is critical and should be well documented.

- Name of analyte (reference standard) and internal standard (if applicable). Add structures and molecular weights as applicable
- Identify if the reference standard is weighed as free form, or as salt (supply correction factors, if any, to obtain the appropriate form being quantified)
- Certificate of analysis or information on lot number, purity or concentration, moisture content, manufacturer/supplier, expiration/retest date, and storage conditions.

3. **Materials and Supplies**

Reagent and supplies can influence method performance. This is especially critical for studies with analytes that exhibit non-specific binding to certain supplies, or certain materials that may produce an interference with the assay. Documentation of the following elements will aid in method troubleshooting when issues are discovered in the assay:

- Suppliers and grades of chemicals and biological reagents
- Source and quality of water
- Source and description of the materials used in the laboratory such as test tubes, extraction blocks, titer plates, injection blocks, capmats, vials, inserts for autosamplers, etc.

4. **Instruments**

Instruments used for the study should be uniquely identified and documented. These include HPLC–MS/MS systems, plate readers, electrochemiluminescence (ECL) platforms, automated liquid handlers, pipettes, centrifuges, evaporators, etc. For multi-component systems, document all components within the system or refer to an identifier established in the laboratory for the component systems. The following information should be documented for the instruments in use:

- Instrument, model number, manufacturer, and serial number.

5. **Assay Settings**

For Chromatographic Assays document:

- Mobile phases
- Flow rates, gradients, pressures, etc.
- Column (serial no.) and pre-column
- Temperatures set on instruments (include column temp)
- Detector (e.g., mass spectrometer) settings
- Major software names and version numbers used for analysis or processing of data (e.g., Analyst® and Watson®).

For Ligand Binding Assays:

- Wash solutions
- Incubation times and temperatures
- Instrument settings
- Detector settings (e.g., absorption wavelength).

6. **Preparation of Reagents/Mobile Phases**

- Actual weights and/or volumes of the components used
- Lot numbers of reagents used.

7. **Preparation of Stock Solutions of Standard Compounds**

- Balance identification
- Calculations for the preparation of stock solutions
- Actual weights taken
- Actual dilution volumes and solvents used

- Storage location, temperature, and expiration dates for standard solutions
- Any precautions in the handling of stock standards.

8. **Preparation of Calibration Standards and Quality Control Samples (QCs)**

Calibration standards are the backbone of a bioanalytical assay, and QCs assess the performance of the assay. Accurate and precise preparation and proper storage are essential in producing a high-quality assay. Therefore, information on the preparation and storage of calibration standards and QCs must be documented.

- Calculations for the preparation of calibration standards and QCs
- Supplier, species, type and lot number of the matrix, anticoagulant (if applicable)
- Preparation procedure, amounts prepared and dispensed in sub-aliquots for storage
- Storage location and storage conditions for calibration standards and QCs
- Ensure all tubes are appropriately labeled before they are stored in temperature controlled freezers.

9. **Sample Processing**

- Reference the analytical method (if available) or write the detailed analytical procedure in the notebook
- Document any deviations or unexpected observations during the extraction.

10. **Pre-study Assay Evaluation**

To evaluate the performance of the system, or to verify the preparation of bulk calibration standards or QCs, it is useful to analyze a set of calibration standards and QCs without including any study samples in the analytical run. This analytical run is called the Pre-study Assay Evaluation (PSAE). The accuracy and precision statistics of the calibration standards and QCs determine the quality of the PSAE run. Documentation of PSAE runs may be useful in the following situations:

- When bulk standards and/or QCs are made, perform a PSAE run to document the qualification of calibration standards and QCs before they are used in analyzing samples
- Run a PSAE when qualifying or evaluating a system, or to train an analyst on the assay.

11. **Sample Storage Information**

- For study sample analysis, document the total number of samples received, freezer number, and location where the samples are stored. The tracking of the sample movement should also be documented.
- For validation experiments, document the location and ingress and egress of validation and stability samples.

- Document temperature monitoring of storage units (e.g., refrigerators and freezers). In modern laboratories this is generally documented by validated electronic monitoring systems. A manual procedure may also be used in laboratories that do not have an electronic monitoring system or in cases of system failure.

12. **Analytical Runs**

Documentation of analytical runs can be sub-categorized as pre-extraction, during-run, and post-run activities. A typical listing of documentation suggested for each sub-category is given below:

- **Pre-extraction documentation**:
  - Identity of samples in the analytical run, type of analytical or validation run, details of experiments included in the run
  - Identity of repeat samples (include reason for repeat)
  - Identify which samples are diluted and the dilution factors, as well as dilution schemes
  - Sample sequence list for a chromatographic run or a block template/map for micro-titer plates
  - Date of sample processing
  - Any deviations from the written method.
- **During-run documentation**:
  - Identifier(s) of the instrument(s) used
  - Acquisition start date
  - Identity and location of folder(s) for the electronic data collection
  - Chromatographic sample queue/tray name (e.g., Watson® or Softmax® file names) including import file name, if any
  - Any deviation or observation during the run (e.g., run stop due to malfunction of instruments).
- **Post-run documentation**:
  After acquisition of the run, the data is reviewed and processed by the analyst or personnel assigned to this task. The data is then peer reviewed by a designated individual or QC section as per laboratory procedures. The data processing and review is usually performed using an analytical system or LIMS (e.g., Watson®) and data tables are generated. The following information should be documented:
  - Did the run meet acceptance criteria and was the run accepted?
  - Unique/abnormal observations (e.g., poor peak shape for a chromatographic run)
  - For chromatographic runs, document if any chromatograms are manually integrated and document the reason
  - Identify samples that would need reanalysis due to analytical reasons and document the reasons

- Any additional observations made for the run (carry over, etc.)
- Any deviation from the written method
- Peer reviewer and date.

13. **Quality Assurance**

During the course of the study, the Quality Assurance Unit may be involved at different stages of the study (e.g., inspection of the study plan, in-process inspection, data audit, report audit, etc.). Documentation of the review and any inspection report and/or response to the inspection report will remain with the Quality Assurance Unit.

14. **Supplementary Information and General Notes**

The notebook should be organized in a way that information is easily accessible and retrievable. A table of contents is helpful to point the reviewer to the appropriate sections quickly. Usually, along with the notebook, other supplemental data needs to be kept with the study. The following is a listing of the type of data that may be included in the study files. Note that depending on the laboratory setup and SOPs, some of this documentation may be compiled and archived centrally.

- Analytical method
- Work list for analytical runs with aliquot information
- Certificate(s) of Analysis
- Printout of any chromatograms with annotations if the required notes are not saved in the raw data
- Verification of data (e.g., any documentation required according to laboratory SOPs to verify review of documents or manually prepared tables)
- Sample handling files
- Any SOP or protocol deviation
- Correspondence with the sponsor, in vivo sites, etc. (e.g., emails needed for reconstruction of the study)
- Any other ancillary data.

4.4 Reports in Regulated Bioanalysis

In bioanalysis there are two major reports: (1) method validation (validation report) and (2) sample analysis from a clinical or nonclinical study (bioanalytical report). Other reports written in bioanalysis are the bioanalytical method and stability reports, which can be written separately or appended to the validation reports. The guidance documents from the FDA [4] and the EMA [6] describe in general terms what is required in the validation and bioanalytical reports, but do not set standards for writing the reports. The reporting information suggested in these guidance documents are summarized in Table 4.1. Having no set standards, bioanalytical practitioners have developed their own styles of writing reports, resulting in reports of various shapes and sizes. Without uniform standard for report writing, regulatory reviews are also cumbersome and there is a risk of omitting essential information [5]. In an attempt to standardize the format of reports, the GBC Documentation team suggested tables of contents for validation and bioanalytical reports (17).

In regulated bioanalysis the bioanalytical method should be validated before analyzing study samples; but are the validation reports required before initiating samples analysis? The Organization for Economic Co-operation and Development (OECD) has recently clarified that there is no regulatory requirement to finalize the validation report for the method that will be used to analyze a GLP study before initiating samples analysis [19]. There is still a regulatory expectation that methods are fully validated before the results of the study are considered valid [19].

4.5 Submission of Data and Reports in Electronic Format

The Food and Drug Administration Safety and Innovation Act (FDASIA) signed into law on July 9, 2012 expanded FDA authorities to safeguard and advance public health. Under this law, the Federal Food Drug and Cosmetic Act (FD&C Act) was amended by adding Sect. 745A(a), requiring regulatory submissions to be in electronic format. To specify how the FDA intends to implement the electronic submission requirements of the amended act, the FDA issued a guidance in 2014 entitled, "Providing regulatory submissions in electronic format—Submissions under Sect. 745A(a) of the Federal Food, Drug and Cosmetic Act" [20]. This guidance [20] requires that certain submissions (INDs, NDAs, ANDAs, and BLAs) to the Center for Drug Evaluation and Research (CDER) or the Center for Biologics Evaluation and Research (CBER) be made electronically in the format specified by FDA. Submissions not made in the required electronic format will not be received or filed, unless a specific exemption from the electronic submission is obtained. FDA did not provide all details on the required standards in this guidance [20] but stated that guidelines on a variety of requirements and formats for electronic submissions will be distributed in several future guidance documents, which will be released after public notice and comments. Sponsors will have at least 24 months (12–18 months for updates) after final publication of the guidance before the requirements of that specific guidance is expected as part of a submission. These guidance documents mandate how the data and reports in regulatory filings are submitted to the FDA. The impact of these guidances on bioanalytical data and reports is discussed below.

4.5.1 Data Submission

The FDA released the final guidance on providing standardized study data in electronic format in December 2014 [21]. A few months later, in 2015, it released the accompanying study data technical conformance guide [22]. The FDA required that the data contained in electronic submissions should be in a format that the agency can process, review, and archive. The required standards are specified in the Data Standards Catalog available on the FDA website [23]. The Standards Catalog provides a list of currently supported and required standards, Controlled Terminologies (CTs), dates for their implementation, and other pertinent technical information. The online Catalog [23] should be continually monitored to ensure utilization of the up-to-date versions of standards and CTs. The agency may refuse to file or receive electronic regulatory filings that do not conform to the standards specified in the Catalog [23].

The FDA has specified that data exchange standards developed by Clinical Data Interchange Standards Consortium (CDISC) be used for submitting data. CDISC created SDTM (Study Data Tabulation Model) to define the standard structure of the data tables for submission [24]. The model is built around the concept of collecting observations in a study. Observations are normally collected in a series of domains. A domain is defined as a collection of logically related observations within a topic, and is designated by a unique two-character code, e.g., PC, PP, etc. The observations consist of discrete pieces of information collected under named 'variables'. All datasets are structured as flat files with rows representing the observations and columns representing variables.

The SDTM describes the basic concepts and general structures. It is designed to accommodate a broad range of clinical and nonclinical studies data in standardized format. Implementation guides (IGs) provide specific recommendations on how to utilize SDTM for the clinical or nonclinical studies and are available on the CDISC website (www.cdisc.org). SDTMIG (Study Data Tabulation Model Implementation Guide) is used for clinical studies [24], and SENDIG (Standard for Exchange of Non-Clinical Data Implementation Guide) is used for non-clinical studies [25]. Variables, specifications, and domains for collecting data are described in IGs. IGs also describe basic assumptions and business rules, and provide numerous examples for mapping data to the standard format.

Bioanalytical data is only a small subset of the overall data submitted in standardized format. Required bioanalytical data is mostly entered in the Pharmacokinetics Concentrations domain—PC. There are about 40 variables listed for PC in STDMIG and SENDIG. So far relatively few bioanalytical-related variables are described in the PC domain. The current bioanalytical-related variables correspond to sample concentrations, measurement units, sample identification, and test identification. Method performance or validation parameter variables do not yet appear in these models. But this may change in the future. The FDA recently announced it intends to release two additional guidance documents: (1) Providing Regulatory Submissions in Electronic Format—Bioanalytical Methods Data Standards and (2) Providing Regulatory Submissions in Electronic Format—Standardized Bioanalytical Data. It is anticipated that these guidance documents will further standardize the submission of electronic bioanalytical data.

The study data technical conformance guide [23] recommends that the ideal time to implement SDTM and SEND standards is prior to conducting the study, so that the results presented in the study can be traced back to the original data. While this recommendation is ideal, documenting bioanalytical work with legacy bioanalytical methods is challenging and would require careful planning. The bioanalytical assays used for analysis of clinical or nonclinical studies may have existed long before the SEND/SDTM was introduced. Therefore, the original naming convention for tests and analytes may not be compatible with the terminology suggested in the SDTM. Also, in practice, the analyte names change during the life cycle of the assay. The analytes could be named with a number assigned by the particular sponsor, IUPAC name, short names, generic name, etc. The unknown metabolites are typically named as M1, M2, etc., before they are identified and given a name. With changing of the analyte name, the test name variable in PC may also change. The existing analytical software and LIMS systems are also not fully compatible with the rapidly changing data exchange standards and controlled terminology. With this reality in bioanalytical naming conventions, and the urgency of complying with the upcoming FDA mandate for submitting data in the specified format, one cannot wait until all systems are compatible with SDTM. As a stop-gap measure, some software vendors have substituted the available terms in the report with an SDTM-compatible term. This stop-gap measure may be needed until such time that the systems are fully compatible with the data exchange standards. For the future, it would be desirable for software vendors to include controlled terminology into their software and systems, e.g., analytical instrument processing software or Watson® LIMS.

4.5.2 Bioanalytical Reports in Common Technical Document

The bioanalytical reports (validation and analysis) are written for submission to regulatory agencies. One should adhere to the requirements provided in the Common Technical Document (CTD) from ICH for submission of these reports. CTD was defined by the ICH Multidisciplinary Group 4 Expert Working Group (ICH M4 EWG), and was released by ICH in July 2003 [26]. Upon release, CTD became the mandatory or highly recommended process for submission of new drug applications to the regulatory agencies in the European Union, Japan, and US. CTD is an interface for industry to agency transfer of regulatory information.

CTD is composed of five (5) modules. Module 1 is region specific and modules 2–5 are common for all regions. CTD defines the paper submission and the electronic submission is defined by eCTD. The documentation and formatting requirements provided in the CTD guidance apply to all regulatory reports including bioanalytical reports. Therefore, the requirements given in the ICH CTD guidance should be carefully reviewed and applied for bioanalytical reports [27, 28].

On May 5, 2015 the US FDA published a binding guidance on providing submissions in electronic format for certain human pharmaceutical products and other related submissions using eCTD specifications [29]. This makes it mandatory to submit the NDAs, ANDAs, BLAs, and master files electronically to CDER and CBER after May 5, 2017, 24 months after release of the eCTD guidance [29]. The INDs must be submitted in eCTD format 36 months after the release of the guidance, i.e., May 5, 2018 [29]. Submissions not made in eCTD format will not be filed or received, unless an exception from the electronic submission requirement is obtained. The current guidance and supportive documents to comply with the mandatory requirements can be found on the FDA's eCTD [30] and ICH [26] websites. Make sure that you are using the latest available guidance. ICH CTD documents [27, 28] provide guidance on where summaries and full bioanalytical method validation and sample analysis reports should be placed within the CTD, and also provide information on what is expected in the summaries of bioanalytical methods. All bioanalytical reports and summaries should be connected through hyperlinks for easy access and reference. Since the nonclinical and clinical studies are written in different modules and sections of the CTD, they are discussed below separately.

4.5.3 Nonclinical Studies

The ICH CTD guidance M4S(R2) [27] requires that summaries and full bioanalytical method validation and sample analysis reports are submitted in the filing. Summaries of bioanalytical method validations used in nonclinical studies should be written in module 2 of the CTD. Summaries of bioanalytical methods are written in Sect. 2.6.4.2, located within the Pharmacokinetic Written Summary Sect. 2.6.4. The following assay validation parameters should be described in the bioanalytical method summaries:

- Detection procedures
- Quantification limits
- Performance (e.g., accuracy and precision) during method validation
- Stability of analytes in biological samples.

A summary table of all bioanalytical method validations containing above parameters should be placed in summary Sect. 2.6.4.2. The bioanalytical methods summary section is also an appropriate place to describe the history of method improvements and validations for the compound [16]. It is very useful to include a summary table of all nonclinical studies and identification of the bioanalytical method(s) used to analyze the samples from each study. It is desirable to provide a composite table of QC performance data observed during the analyses.

The full reports describing bioanalytical methods and their validations should be placed in module 4 of CTD, in sub-sect. 4.2.2.1. However, if the nonclinical study report includes the bioanalytical method and validation, the report should be placed in the specified location where the nonclinical report resides. Sample analysis reports should be part of the nonclinical study report, and placed in the required location for nonclinical study reports..

4.5.4 Clinical Studies

The ICH CTD guidance M4E(R1) [28] requires that summaries and full bioanalytical method validation and sample analysis reports are provided in the regulatory filing. An overview or summary of bioanalytical methods and their validation should be written in module 2 of CTD, sub-sect. 2.7.1 (Summary of Biopharmaceutic Studies and Associated Analytical Methods). The following assay validation parameters should be described within the bioanalytical method summaries:

- Detection procedures
- Linearity range, sensitivity and specificity
- Assay performance (e.g., accuracy and precision) during methods validation
- Stability of analytes in biological samples.

A summary table of all bioanalytical method validations containing the above parameters should be placed in summary Sect. 2.7.1. As described under non-clinical studies, the history of method improvements and reasons for multiple validations (if any) for the compound should be provided in the summary section [16]. It is very useful to include a summary table of all clinical studies and identification of the bioanalytical method(s) used to analyze the samples from each study. It is desirable to provide a composite table of QC performance data observed during the analyses.

The full reports describing bioanalytical methods and their validations should be placed in module 5 of CTD, Sect. 5.3.1.4, and referenced in individual clinical reports. If the clinical report contains a unique method which is used only once for the particular clinical study, the bioanalytical method and validation report should be placed within the clinical study report. The sample analysis reports should be part of the clinical study report, and placed in the required location for clinical study reports.

Acknowledgements The author would like to thank Jacueline Gillis, Gerard Randolph, and Herbert Birnboeck for their valuable edits and suggestions.

References

1. Shah VP, Midha KK, Dighe SV, et al. Analytical methods validation: bioavailability, bioequivalence and pharmacokinetic studies. Pharm Res. 1992;9(4):588–92.
2. Shah VP, Bansal SK. Historical perspective on the development and evolution of bioanalytical guidance and technology. Bioanalysis. 2011;3(8):823–7.
3. Shah VP, Midha KK, Findlay JW, et al. Bioanalytical method validation—a revisit with a decade of progress. Pharm Res. 2000;17(12):1551–7.
4. US Department of Health and Human Services, Food and Drug Administration, CDER, CVM, guidance for industry: bioanalytical method validation. 2001. http://www.fda.gov/downloads/Drugs/Guidances/ucm070107.pdf. Accessed 10 May 2016.

5. Au S, Yu C, Booth B. Utilization of electronic resources in the NDA/BLA regulatory review of bioanalytical data: perspectives from US FDA reviewers. Bioanalysis. 2011;3(13):1441–5.
6. European Medicines Agency, Committee for Medicinal Products for Human Use. Guideline on validation of bioanalytical methods, [EMEA/CHMP/EWP/192217/2009]. July 2011. http://www.ema.europa.eu/docs/en_GB/document_library/Scientific_guideline/2011/08/WC500109686.pdf. Accessed 10 May 2016.
7. Japan Ministry of Health, Labour and Welfare (MHLW), guideline on bioanalytical method validation in pharmaceutical development. 2013. http://www.nihs.go.jp/drug/BMV/250913_BMV-GL_E.pdf. Accessed 10 May 2016.
8. Japan Ministry of Health, Labour and Welfare (MHLW), guideline on bioanalytical method (Ligand Binding Assay) validation in pharmaceutical development. 2014. http://www.nihs.go.jp/drug/BMV/260530_LBA-GL_E.pdf. Accessed 10 May 2016.
9. China State Food and Drug Administration. Technical guideline on clinical pharmacokinetic studies for chemical drugs. China: Beijing; 2005.
10. Chinese Pharmacopoeia 2015, Guidance on Bioanalytical Method Validation, Vol. 4, pp. 363–8.
11. Agência Nacional de Vigilância Sanitária (ANVISA), Brazil, Resolution RDC27 (Bioanalytical Guidance). 2012. http://bvsms.saude.gov.br/bvs/saudelegis/anvisa/2012/rdc0027_17_05_2012.pdf. Accessed 10 May 2016.
12. Ministry of Health, Health Products and Food Branch, guidance document: conduct and analysis of comparative bioavailability studies, Canada, 2012. http://www.hc-sc.gc.ca/dhp-mps/alt_formats/pdf/prodpharma/applic-demande/guide-ld/bio/gd_cbs_ebc_ld-eng.pdf. Accessed 10 May 2016.
13. Bansal SK. International Regulations and Quality Standards of Bioanalysis. In: Li W, Zhang J, Tse FLS, editors. Handbook of LC-MS bioanalysis: best practices, experimental protocols, and regulations. New York: Wiley; 2013.
14. U.S. Department of Health and Human Services, Food and Drug Administration, guidance for industry: bioanalytical method validation, draft guidance, 2013. http://www.fda.gov/downloads/drugs/guidancecomplianceregulatoryinformation/guidances/ucm368107.pdf. Accessed 10 May 2016.
15. Viswanathan CT, Bansal S, Booth B. Quantitative bioanalytical methods validation and implementation: best practices for chromatographic and ligand binding assays. Pharm Res. 2007;24(10):1962–73.
16. Booth B, Arnold, ME DeSilva B, Amaravadi L, Dudal S, Fluhler E, Gorovits B, Haidar SH, Kadavil J, Lowes S, Nicholson R, Rock M, Skelly M, Stevenson L, Subramaniam S, Weiner R, and Woolf E. Workshop report: crystal city V—quantitative bioanalytical method validation and implementation: The 2013 Revised FDA Guidance. AAPS J. 2015;17(2).
17. Verhaeghe T, Barton HH, Hara H, Hucker R, Kelley M, Picard F, Reddy KS, Salvadori MC, Woolf E. Recommendations from the global bioanalysis consortium team A8: documentation. AAPS J. 2014;16(2).
18. United States Pharmacopeia, Chapter <1058>, Analytical Instrument Qualification, Rockville, USA.
19. OECD Good Laboratory Practice: Frequently Asked Questions (FAQ) posted on Jan. 16, 2016. Available at http://www.oecd.org/env/ehs/testing/glp-frequently-asked-questions.htm. Accessed 10 May 2016.
20. US Department of Health and Human Services, Food and Drug Administration, Providing Regulatory Submissions in Electronic Format—Submissions Under Section 745A(a) of the Federal Food, Drug and Cosmetic Act, Guidance for Industry, FDA. December 2014. http://www.fda.gov/downloads/Drugs/Guidances/UCM384686.pdf. Accessed 10 May 2016.
21. US Department of Health and Human Services, Food and Drug Administration, Providing rRegulatory Submission in Electronic Format—Standardized Study Data, Guidance for Industry, FDA. December 2014. http://www.fda.gov/downloads/Drugs/Guidances/UCM292334.pdf. Accessed 10 May 2016.

22. US Department of Health and Human Services, Food and Drug Administration, Study Data Technical Conformance Guide, Technical Specifications Document. 2015. http://www.fda.gov/downloads/ForIndustry/DataStandards/StudyDataStandards/UCM384744.pdf. Accessed 10 May 2016.
23. US Food and Drug Administration, Data Standards Catalog v4.4 (2015) available in study data standards resources http://www.fda.gov/forindustry/datastandards/studydatastandards/default.htm. Accessed 10 May 2016.
24. CDISC Submission Data Standards Team, Clinical Data Interchange Standards Consortium, Study Data Tabulation Model (SDTM). http://www.cdisc.org/sdtm. Accessed 10 May 2016.
25. CDISC Standard for Exchange of Nonclinical Data Team, Clinical Data Interchange Standards Consortium, Standard for Exchange of Nonclinical Data (SEND) implementation guide: nonclinical studies, v. 3.0. 2011. http://www.cdisc.org/send. Accessed 10 May 2016.
26. The International Council for Harmonisation, M4: the common technical document. http://www.ich.org/products/ctd.html. Accessed 10 May 2016.
27. The International Council for Harmonisation, The common technical document for the registration of pharmaceuticals for human use: safety—M4S(R2), nonclinical overview and nonclinical summaries of module 2, organization of module 4. 2002. http://www.ich.org/fileadmin/Public_Web_Site/ICH_Products/CTD/M4__R2__Safety/M4S_R2_.pdf. Accessed 10 May 2016.
28. The International Council for Harmonisation (ICH), the common technical document for the registration of pharmaceuticals for human use efficacy—M4E(R1), clinical overview and clinical summary of module 2, module 5: clinical study reports, 2002. http://www.ich.org/fileadmin/Public_Web_Site/ICH_Products/CTD/M4__R1__Efficacy/M4E__R1_.pdf. Accessed 10 May 2016.
29. US Department of Health and Human Services, Food and Drug Administration, Providing Regulatory Submissions in Electronic Format—certain human pharmaceutical product applications and related submissions using the eCTD specifications, Guidance for Industry. May 2015. http://www.fda.gov/ucm/groups/fdagov-public/@fdagov-drugs-gen/documents/document/ucm333969.pdf. Accessed 10 May 2016.
30. US Food and Drug Administration, Electronic Common Technical Document (eCTD) website. http://www.fda.gov/Drugs/DevelopmentApprovalProcess/FormsSubmissionRequirements/ElectronicSubmissions/ucm153574.htm. Accessed 10 May 2016.

Chapter 5
Regulated Bioanalysis—Historical Aspects and Key Concepts Related to Its Use

Sheila Breidinger and Eric Woolf

Abstract The history and key concepts of bioanalysis are discussed. Historically, the scientific need driving the development of bioanalytical methodology was the evolution of the science of pharmacology as well as the field of pharmacokinetics. Similarly, the ability to actually perform bioanalysis was linked to the development of analytical methodology/instrumentation. Each of these historical aspects of bioanalysis is discussed in this chapter. Key attributes of molecules quantitated in bioanalytical procedures for small molecules and protein therapeutics are described. Topics discussed in this regard include plasma protein binding, free drug versus total drug measurements, prodrugs versus active moieties, metabolites, minimum required dilution, and parallelism. In addition, important considerations with respect to different types of study samples are described. These factors include the influence of dosing vehicle and collection methodology. Matrices discussed include whole blood, urine, cerebral spinal fluid, tissues, and dried blood spots.

Keywords History of bioanalysis · Sample matrix · Plasma protein binding · Dosing vehicle effects · Prodrug · Whole blood · Urine · CSF · Tissues · Dried blood spot

5.1 A Brief History of Bioanalysis

The history of bioanalysis is tied directly to the development of the science of pharmacology as well as the field of pharmacokinetics. Practical bioanalysis is linked to the development of analytical methodology and instrumentation to support the data needs of these fields. In this chapter, we discuss the inter-relationships and histories of these fields.

S. Breidinger · E. Woolf (✉)
Merck Research Labs—PPDM Clinical Bioanalytics, West Point, PA 19486, USA
e-mail: eric_woolf@merck.com

S. Breidinger
e-mail: sheila_breidinger@merck.com

M.L. Rocci Jr. and S. Lowes (eds.), *Regulated Bioanalysis: Fundamentals and Practice*, AAPS Advances in the Pharmaceutical Sciences Series 26,
DOI 10.1007/978-3-319-54802-9_5

In 1847, during a presentation to the Philosophical Society of Glasgow, Andrew Buchanan, M.D. postulated that "ether only acts as a narcotic after being absorbed, and that the energy of its action is proportionate to the degree in which the blood applied to the tissues of the heart and brain is impregnated with it" [1]. This presentation is one of the first reported instances relating blood concentration to an observed pharmacologic action [2], though almost 170 years ago, the technology to confirm Buchanan's hypothesis did not exist.

Colorimetric methods were the earliest approach utilized for bioanalysis. These methods rely on the formation of a colored complex between the target analyte and a reagent. According to Beer's Law, analyte concentration is directly proportional to the absorbance of the colored complex, hence concentration can be determined based on the intensity of the colored complex as determined using a colorimeter. The specificity of colorimetric methods is dependent upon the number of substances present in the sample that can form the colored complex with the chosen reagent. An example of one such method is the "System of Blood Analysis" described by Folin and Wu for the determination of nonprotein nitrogen, urea, creatinine, creatine, uric acid, and sugar in blood [3]. It is worth noting that such colorimetric methods were not limited to endogenous components, a point exemplified by publications of Haggard [4] and Shaffer [5] in 1923 describing methods for the determination of diethyl ether in blood.

A major issue with such colorimetric methods is lack of specificity. Any compound that reacts with the colorimetric reagent to form a complex will yield a positive result. In other words, the degree of color formation, and hence analyte concentration, is dependent upon the total concentration of all analytes in the sample capable of reacting with the reagent. During the 1940s and 1950s, the lack of specificity was addressed by research discoveries in two areas: chromatography and immunoassay. Seminal works in both these fields were the subject of Noble Prizes, with A.J.P. Martin receiving a Noble Prize for his research in chromatography in 1952, and Rosalyn Yalow receiving a Nobel Prize in 1977 for her research in the area of radioimmunoassay.

Chromatographic methods improved assay specificity by adding a separation "dimension" prior to analyte detection. The earliest chromatography-based bioanalytical methods utilized gas chromatography (GC) for separation, prior to a non-specific detector such as a flame ionization detector (FID). Additional specificity for chromatographic-based assays could be gained by using a mass-specific detector, i.e., a mass spectrometer.

A prerequisite for the application of GC is the need for the analyte to be volatile, hence limiting direct application of GC-separated analytes for detection. In order to overcome this requirement, researchers developed novel methods to derivatize analytes to volatile species. In these methods, the analyte is typically first extracted from the matrix and then treated with a reagent that reacts with a functional group on the analyte to form a volatile species. Excess reagent is then removed from the sample, and the sample is injected into a gas chromatograph using conditions that separate the analyte from the other components in the sample. The derivatized analyte is subsequently detected using either a universal detector (e.g., FID), or a specific detector such as an electron capture detector, nitrogen–phosphorus GC detector, or a mass spectrometer; the latter having the advantage that it can confirm the structural integrity of the derivatized analyte.

The limited applicability of GC-based methods contributed to rapid adoption, during the 1970s, of bioanalytical methods based on high performance liquid chromatography (HPLC), which removes the requirement for an analyte to be volatile. In contrast to GC, the major limitation of early HPLC methods was sensitivity. The most commonly employed HPLC detection approach in early methods was based on ultraviolet/visible (UV/Vis) spectrometry. Hence, sensitivity was directly dependent upon the extinction coefficient of the analyte at the detection wavelength employed. In order to address this sensitivity issue, researchers developed derivatization approaches to introduce chromophores to analytes. Other detection methods based, for example, on fluorescence or electrochemistry were also employed.

The sensitivity issue for HPLC was largely addressed when the atmospheric pressure HPLC/mass spectrometry interface was introduced in the early 1990s. Today, LC-MS is the most commonly used analytical platform for small molecule bioanalysis. The shift from GC-based bioanalytical methods to LC-MS methods is illustrated by the subjects of presentations at the Reid International Bioanalytical Forum, held every 2 years since it was initiated in 1975. According to Stevenson's review [6], during the first three meetings, the number of presentations describing GC-based methodology outnumbered those describing HPLC methods by a ratio of 2:1. During the 1980s and 1990s, this ratio gradually shifted to favor HPLC assays. HPLC-MS methods began to predominate in 1997, and since then the majority of the methodology presentations at this conference have been HPLC-MS based.

Immunoassay-based methodology developed in parallel to chromatographic methods through the 1980s and 1990s. Radioimmunoassay specificity was initially based upon the three-way interaction between the analyte, radioactive analyte, and an antibody to the analyte. The downside of such methods, however, was the need for appropriate reagents; that is, a specific antibody for the analyte in question as well as the labeled analyte upon which detection is based. The advantage of such methods, especially those utilizing radioactivity detection, is the potentially high degree of sensitivity that such methods afford.

During the 1970s and early 1980s, immunoassay-based methods filled, at least in part, the sensitivity gap experienced with chromatographic assays. As a result, many new drug applications (NDAs) during this period employed bioanalytical methods that were immunoassay based. For example, the NDAs for enalapril and lisinopril, two of the most commonly prescribed drugs for hypertension, were supported entirely with immunoassays [7].

Presently, LC-MS based methodology has largely eliminated the need to develop immunoassay methods to support small molecule (typically less than 1000 g/mol molecular weight) bioanalysis. On the other hand, immunoassay methods are still widely employed for the measurement of proteins and peptides in biological fluids.

5.2 Basic Principles of Pharmacokinetics

A major application of bioanalytical data is its use in elucidating a molecule's pharmacokinetics. The word pharmacokinetics is derived from the Greek words pharmakon, meaning drugs and poisons, and the word kinetics, referring to the change of one or more variables as a function of time. The purpose of pharmacokinetics is to study the time course of drug or metabolite concentrations or amounts in biological matrices and to construct suitable models for interpreting such data [2]. The term first appeared in 1953 in a publication by Dost [2].

Some basic pharmacokinetic terms derived from bioanalytical data, along with their definitions, are as follows.

Area under the curve: The integrated area under the molecule's plasma concentration–time curve between times t1 and t2 (Note: depending on the molecule, serum, or whole blood concentrations of the drug may be evaluated over time, as opposed to plasma concentrations—for the purposes of this section we will consider plasma, serum, and whole blood to be interchangeable). This parameter is represented mathematically by the equation $\int_{t1}^{t2} Cdt$. The term is typically abbreviated $AUC_{t1\text{-}t2}$ and is expressed in units of (ug/L) · hr or uM · hr.

Half-life: Time required for drug plasma concentrations to decrease by one-half. Elimination half-life generally refers to the half-life observed in the terminal portion of the concentration–time curve. The term is typically abbreviated $t_{1/2}$.

Maximum Concentration: The maximum plasma concentration of drug observed following its administration. This is typically abbreviated C_{max}.

Time of Maximum Concentration: The time after drug administration at which C_{max} is reached. This term is typically abbreviated T_{max}.

Volume of Distribution: The apparent volume in the body in which a drug is distributed. Typically abbreviated V_d, this value is calculated as the dose, D, divided by the concentration of drug at $T = 0$, C_0. Because blood volume in humans is approximately 5L, a volume of distribution significantly larger than 5L indicates that the compound likely distributes into extravascular space.

Plasma Clearance: Volume of plasma cleared of the drug per unit time. This value is calculated by dividing the systemically available dose (see below) by total AUC extrapolated to time infinity.

Two additional key terms that are of particular relevance to regulated bioanalysis are bioavailability and bioequivalence.

Bioavailability: Typically symbolized by a lower case f, bioavailability refers to the fraction of a dose of drug that reaches the systemic circulation and is generally characterized as either relative (f_{rel}) or absolute (f_{abs}).

- *Relative bioavailability* refers to a comparison of, or ratio of, drug available systemically after the administration of equivalent doses of two different, generally oral, dosage forms. This value is calculated by taking the ratio of the AUCs obtained after the two different dosage forms are administered.

- *Absolute bioavailability* refers to a comparison of the amount of drug systemically available after its administration through a route where the drug needs to be absorbed (typically oral) to that following intravenous administration. Because it is often not feasible to administer the same dose via both IV and oral routes, it is necessary to add a dose correction factor to f_{abs} calculations. Hence $f_{abs} = (AUC_{po} \cdot D_{iv})/(AUC_{iv} \cdot D_{po})$, where AUC_{po} and AUC_{iv} are the areas under the curve obtained following oral and IV administration, respectively, while D_{iv} and D_{po} are the respective IV and orally administered doses.

Bioequivalence is defined by the US FDA as "the absence of a significant difference in the rate and extent to which the active ingredient or active moiety in pharmaceutical equivalents or pharmaceutical alternatives becomes available at the site of drug action when administered at the same molar dose under similar conditions in an appropriately designed study" [8]. In other words, the amount of drug present systemically is similar when two different but bioequivalent dosage forms of a drug are administered.

From a practical perspective, bioequivalence between dosage forms is assessed by conducting a study in which the dosage forms being compared are administered to study participants, after which blood samples from the participants are obtained and analyzed in a regulated bioanalytical setting. In order to assess bioequivalence, the AUCs of the plasma concentration–time curves obtained after each formulation is administered are calculated. In addition, the C_{max} for each treatment is reviewed. The formulations are generally considered bioequivalent if the 90% confidence intervals of the ratios of the means of the AUC and C_{max} for the formulations lie in the range of 80–125%.

5.3 Regulated Bioanalysis in the US—A Historical Perspective

The regulations and guidance with respect to the performance of bioanalysis are covered elsewhere in this book. In contrast, the history behind the requirement to include bioanalytical data in applications to market new drugs in the United States is not widely known and is also appropriate to include in a text dealing with the practice of regulated bioanalysis. For additional details, we refer the reader to a review article by Skelly [9] covering the history of biopharmaceutics in the Food and Drug Administration.

The requirement to include bioanalytical data in new drug applications in the United States can be traced back to the 1962 amendment of the Food, Drug, and Cosmetic Act (FDC). Prior to this revision, sponsors of new drug applications needed only to demonstrate that the subject product of the application was safe. The new law required applicants to demonstrate evidence of efficacy in addition to toxicological data before a new drug could be marketed.

A component of the 1962 amendment was the requirement, applicable to drugs already marketed and for which efficacy data was lacking, for sponsors to demonstrate that the dosage form then in the marketplace was indeed absorbed into the body. In addition, around this time, there was concern that the bioavailability of different formulations of generic drugs then on the market was not equivalent to the initially marketed brand [9]. The US FDA interpreted the 1962 amendments as a mandate that all generic drug formulations marketed between 1938 and 1962 must establish equivalency to the approved marketed brand.

This interpretation led to litigation on FDA's authority to impose such requirements. Eventually, however, the US Supreme Court ruled that it was in the interest of the American people that FDA assures the equivalency of generic drug products, and upheld FDA's authority to require this information.

Following the court decision, the FDA developed and published draft regulations applicable to both innovator and generic drug products. These regulations were finalized, published in the US Code of Federal Regulations (CFR), and became effective in 1977. Regulations in 21 CFR part 320 establish definitions and requirements for bioavailability and bioequivalence studies. Specifically, 21 CFR part 320.29 requires "The analytical method used in an in vivo bioavailability or bioequivalence study to measure the concentration of the active drug ingredient or therapeutic moiety, or its active metabolite(s), in body fluids or excretory products, or the method used to measure an acute pharmacological effect shall be demonstrated to be accurate and of sufficient sensitivity to measure, with appropriate precision, the actual concentration of the active drug ingredient or therapeutic moiety, or its active metabolite(s), achieved in the body," thus ushering in the age of regulated bioanalysis in the United States.

5.4 Validation

The legal requirement to ensure that bioanalytical methods yield results that are accurate and precise led to the need to develop validation criteria for such methods. Validation is defined as "establishing documented evidence that provides a high degree of assurance that a specific process will consistently produce a product meeting its predetermined specifications and quality attributes" [10]. In the case of bioanalysis, the process is the bioanalytical method and the products are the analytical results generated by the method.

Validation is a means to assess assay performance to determine, in advance of use, whether the assay is capable of generating analytical results that meet regulatory requirements. The goal of validation is to determine, a priori, the impact of various situations that may be encountered during sample analysis. Performance characteristics typically evaluated during assay validation include both intra- and inter-run precision and accuracy, stability of the analyte both in working solutions and in matrix, assay specificity/selectivity, extraction recovery, and impact of matrix effects on assay accuracy and precision. Depending upon the conditions to which samples may be exposed, other assessments may need to be conducted. For example, freeze-thaw stability assessment is necessary if samples are frozen prior to analysis. Additional studies may be required when immunoassay techniques are validated to support the development of protein therapeutics (see Sect. 5.6 below for more information).

Acceptable assay performance is demonstrated when an assay, during validation, meets previously defined acceptance criteria. The criteria are generally defined in either a standard operating procedure or validation plan. These criteria are typically based on regulatory guidance described elsewhere in this book.

5.5 Attributes of Small Molecule Drug Bioanalysis

Having described various aspects of the scientific and regulatory history of bioanalysis, we now turn attention to some key attributes of the molecules quantitated in bioanalytical procedures.

5.5.1 Plasma Protein Binding Free Versus Total Measurements

Plasma protein binding (PPB) is a measure of the fraction of drug bound to proteins [i.e., albumin, α1-acid glycoprotein (AGP), globulins and lipoproteins] in plasma. The binding is generally reversible and equilibrium exists between bound and free drug. Protein binding is often evaluated early in drug discovery to characterize drug candidates. Unbound drug is generally the pharmacologically active component and therefore PPB is an important measure necessary for PK/PD predictions.

Many different methods can be used to measure PPB. Separative methods like equilibrium dialysis, ultrafiltration, ultracentrifugation, parallel artificial membrane permeability assay (PAMPA), and capillary electrophoresis (CE) use physical means to separate protein bound and free drug prior to bioanalysis. Non-separative methods like spectrometric and calorimetric techniques that measure changes in physiochemical properties as a function of protein binding have also been used to measure PPB [11].

Equilibrium dialysis is generally considered the gold standard and uses a two compartment chamber with a semipermeable membrane separating the chambers. The pore size of the membrane is selected to permit unbound drug to pass from the chamber containing the biological sample to one containing physiologic buffer when incubated, agitated, and allowed to reach equilibrium. Drug concentrations on both sides of the membrane are then determined to calculate the extent of protein binding [12].

5.5.2 Prodrug Versus Active Moiety

Prodrugs are generally inactive precursors of active drug moieties. Metabolism of a prodrug in vivo can convert it to the pharmacologically active form. Prodrugs are often used to increase the bioavailability of an active drug moiety. A prodrug is generally cleaved quickly to the active drug form, typically at higher levels than would be achieved after the direct administration of the active moiety. From a bioanalytical perspective, it is necessary to ensure stability of both the drug and prodrug during sample collection, storage, and extraction. If samples are not adequately stabilized, additional conversion can occur post collection and lead to erroneously high drug concentrations upon sample analysis.

Ester prodrugs are one of the most common forms of prodrugs. Ester bonds can be hydrolyzed by esterases which are inherently present in biological matrices and can lead to ex vivo conversion of a prodrug if a sample is not stabilized adequately. Use of a screening approach to select the best additive to ensure stability of ester-containing prodrugs was the subject of a recent publication by Fung [13].

Testing must be done as part of method development and validation to ensure post collection prodrug stability through the appropriate choice of matrix (i.e., plasma or serum), anticoagulant, and additives. Use of cold temperature (−80 °C) sample storage, sample extraction on ice, and addition of an acid and/or an enzymatic additive are approaches typically used to stabilize samples containing prodrugs. Examples of some enzymatic additives that have been used to stabilize ester-containing prodrugs include: Sodium Fluoride (NaF), paraoxon, phenylmethanesulfonyl fluoride (PMSF), bis(4-nitrophenyl)-phosphate (BNPP), acetylcholine, and dichlorvos [13]. The use of hazardous chemicals like dichlorvos at the clinic should be avoided if there are safer alternatives.

In practical terms, prior to validation, quality control samples (QCs) containing only prodrug should be prepared in fresh biological matrix. They should be monitored from collection through extraction in order to assess conversion of the prodrug to the active drug. Minimizing exposure to ambient temperature, limiting freeze-thaw cycles, addition of stabilizers (acid, esterases), are all practical ways to achieve incurred sample stability.

5.5.3 Metabolites

Chromatographic resolution of metabolites from the parent drug is important due to the potential for error-inducing interferences particular to study samples (i.e., not occurring in spiked calibrators or QCs). An LC-MS system's ionization source can produce events that lead to such interferences. N-oxide and conjugated metabolites can be converted back to their parent in the APCI or ESI source [14]. Although the extent of reactivity is compound dependent, the analyst should investigate for interference while setting up chromatography and detection parameters. Metabolites may also need to be stabilized at the point of sample collection in order to avoid back conversion to the parent drug. One example is an acyl glucuronide metabolite, which can be hydrolyzed to parent drug or is susceptible to acyl migration. Additionally, lactones can undergo conversion between the closed and open ring forms in a reversible way. These conversions can be minimized through acidification of the sample to carefully control the pH [15]. However, addition of acid in solution can dilute the sample and must be considered when assaying the samples. Also, addition of too much acid (even inadvertently at the clinic by a technician unfamiliar with the process) can denature plasma proteins resulting in sample gelling that could render the plasma sample unusable.

5.6 Attributes of Assays for Protein Therapeutics

5.6.1 *Free Versus Total Analysis*

In contrast to small molecule therapeutics, which generally bind to plasma proteins nonselectively, therapeutic antibodies and other protein therapeutics are typically designed to form specific, non-covalent, high-affinity bonds to protein or peptide targets (i.e., antigens) in order to elicit a pharmacologic response. Antibodies, based on their structure, generally have two independent antigen binding sites. Many antibody targets are soluble proteins or peptides which are in systemic circulation. Hence, in such cases, antibodies in circulation may be found unbound to their target (i.e., free), partially bound to their target (i.e., one of two antigen binding sites occupied), or fully bound (i.e., both binding sites occupied). The relationship between bound and free antibody under equilibrium conditions is described by the equation:

K = (molar concentration of the antigen–antibody complex)/[(molar concentration of the unbound antigen binding sites)(molar concentration of the unbound antibody binding sites)], where K is the equilibrium constant.

Bioanalytical methods for protein therapeutics, ligand binding assays in particular, can be designed to measure either free or total analyte. The selection of which form to measure, as well as approaches to the measurement of each form, has been the topic of much discussion in the literature and is beyond the scope of this chapter. For additional information on this topic, we refer the reader to a review paper by Kuang et al. [16].

5.6.2 *Minimum Required Dilution*

In order to maximize sensitivity and minimize matrix interferences, samples for ligand binding assays are generally diluted prior to analysis. “The minimum required dilution (MRD) for an assay is the minimum magnitude of dilution to which a sample must be subjected to optimize accuracy and precision in an assay run with a specified standard and sample diluent" [17].

The MRD for an assay should be determined during assay development, and acceptable assay performance with respect to sensitivity, precision, accuracy, and analyte response of samples using the MRD should be demonstrated during assay validation.

5.6.3 *Parallelism*

Taking dilution factor into account, any sample dilution greater than the MRD should yield the same back calculated concentration. Parallelism refers to an assessment of the concentrations obtained when study samples are analyzed with multiple different dilution factors greater than the MRD. In general, because post-dose samples are needed for parallelism assessment, this assessment cannot be conducted pre-study.

Parallelism is typically assessed using pooled Cmax samples from a study and analyzing them with gradually increasing dilution factors beyond the MRD. As a target, it is recommended that the relative standard deviation (%CV) between samples in a dilution series be no more than 30% [17].

5.7 Study Sample Matrix Influence and Challenges

5.7.1 *Plasma/Serum—Collection Methodology*

Anticoagulants are used to prevent blood from clotting to preserve the integrity of samples used for bioanalysis. Each new bioanalytical assay involves choosing an anticoagulant. Historically, heparin was the anticoagulant of choice but tended to leave more floating lipids in the plasma and was replaced by EDTA as a favorite when automation became more prevalent [18]. Sample collection tubes containing spray-dried K_2EDTA (lavender-topped tubes) are currently the most commonly employed tubes for samples targeted for bioanalysis; the anticoagulant in these tubes is in a dry format, hence eliminating sample dilution effects. When selecting the anticoagulant and volume of sample collection at the clinic, the practitioner should verify that tubes are available in the desired volume with the additive of choice.

Another point to remember when selecting the anticoagulant is that, depending on the analyte, the anticoagulant may alter the partitioning of the analyte between red blood cells and plasma, thus altering analyte concentration in plasma. For example, it has been reported that for the antibiotic tigecycline, plasma concentrations from samples collected from rats in tubes containing EDTA were two fold lower than those collected in tubes containing heparin as anticoagulant [19]. Similar effects of a lesser magnitude were observed for ciprofloxacin. We hypothesize that the anticoagulant might compete with the analytes for chelating metal ions, thus altering the partitioning of drug between blood and plasma.

While regulatory guidances state that use of a different anticoagulant (i.e., EDTA versus heparin) requires additional validation, there has been much discussion on whether a change to the counterion (sodium, potassium, lithium) should require additional validation. Most counterions are already circulating in biological matrices and may not significantly change the matrix composition. Consequently, in the most recent version of the EMA guidance [20] and FDA draft guidance [21], partial validations for different anticoagulants are mentioned but validation is not required for a change of counter-ion.

5.7.2 Dosing Vehicle Effects

Especially in early discovery PK studies, where drug formulations are often solutions in high concentration of dosing vehicles like polyethylene glycol (PEG) 400 and Tween 80, testing to ensure the dosing vehicle does not cause a matrix effect is critical. If post-dose samples show signal suppression, additional sample cleanup or chromatographic separation may be needed [22]. With LC-MS assays, differences observed between internal standard peak areas in post-dose samples compared to pre-dose or control samples can indicate a matrix effect. This is usually tested by comparing freshly spiked low and high QC samples prepared in control matrix to QCs prepared in matrix spiked with a small percent of dosing vehicle.

5.7.3 Whole Blood

Occasionally whole blood is used as the matrix of choice when an analyte preferentially distributes into red blood cells. This is usually necessary when a drug binds strongly to the red blood cells and the use of plasma as a matrix would result in significantly underestimated concentrations of a drug in circulation. However, working with whole blood in the bioanalytical laboratory has some inherent challenges. Blood is more viscous than plasma, therefore the use of wide bore pipet tips or slower aspiration speed may be necessary to improve pipetting consistency. Also, addition of an organic solvent to ensure red blood cells in samples and control blood are consistently lysed prior to extraction should be explored.

5.7.4 Urine

Urine can be a very challenging matrix to work with due in part to large subject to subject differences in pH, concentration of salts, urine volumes collected, and drug solubility [23]. In addition, nonspecific binding of the analyte to the collection containers should be evaluated in method development prior to sample collection. Bovine serum albumin (BSA) addition can be very effective for analytes that are highly bound to proteins as it adds protein to the urine to make the matrix more like plasma, which contains 6–8% proteins [24]. Alternatively, addition of surfactant or detergents like Tween 20 [25], Tween 80, sodium dodecyl benzene sulfonate (SDBS) [26], and 3-[(3-cholamidopropyl) dimethylammonio]-1-propanesulphonate (CHAPS) [27] are also effective ways to improve urine recovery by limiting adsorption. If the analyte binds to the container, the measured drug concentration in urine will be underestimated, therefore an additive needs to be part of method development prior to the clinical collection procedure.

5.7.5 CSF

Cerebral spinal fluid (CSF) drug levels have been explored as a surrogate measure of brain penetration. The low levels of proteins in CSF can lead to nonspecific adsorption similar to that observed with urine. Recovery of analyte in CSF should be evaluated prior to collection in case vials of different material (i.e., glass, polypropylene) or additives are needed to improve recovery. The type of catheters used to collect CSF at the clinic should also be evaluated if nonspecific binding is an issue, so the best materials can be suggested for use at the clinic [28]. Blood can also contaminate lumbar punctures and potentially lead to inconclusive results.

5.7.6 Tissue Samples

Tissue analysis can provide valuable physiological information during drug development, investigating questions like:

- Is an oncology drug present in tumor tissue?
- Does a central nervous system (CNS) drug cross the blood brain barrier (BBB) into brain tissue?
- Does a transdermal patch absorb into the surrounding skin?
- Does an HCV medication penetrate into the liver?

These inquiries provide examples of how tissue analysis can provide insight into the mechanism of action of a drug. However, tissue analysis poses some significant challenges in the bioanalytical laboratory. Although matrix assisted laser desorption ionization mass spectrometry (MALDI MS), nuclear magnetic resonance (NMR), and quantitative whole-body radiography (QWBA) techniques have been used to measure intact tissue concentrations, tissue homogenate approaches are more common and utilize equipment (LC-MS/MS) already available in a bioanalytical laboratory. Homogenization of tissue can be accomplished using sonication, digestion, or mechanically through grinding, blades, or shaking with beads. [29] After the sample is homogenized, it can be extracted for analysis using standard techniques like protein precipitation (PPT), liquid–liquid extraction (LLE), or solid phase extraction (SPE). Standards and QCs are prepared in control homogenate. One liability in the use of tissue homogenate is that stability in homogenate may not be comparable to stability in pure tissue. It is not possible to truly mimic post-dose tissue when preparing QCs, but one way to mitigate this issue is to collect multiple tissue samples and test incurred sample stability. This assumes that the drug concentration across different tissue samples is homogeneous.

Collection of tissues in animal species is often done during necropsy where the whole tissue or portion of the tissue may be collected for analysis. However, in human studies tissues are usually collected though biopsy sampling. When collecting a tissue sample, care must be taken to ensure fluids surrounding the tissue are not also collected because they could add drug concentration to or dilute the sample. Tissues are often rinsed prior to placement in a vial and then frozen for shipment to the bioanalytical lab. Practical advice on how to prepare tissues for bioanalysis is offered in the review article by Kemper et al. [30].

5.7.7 Dried Blood Spots

An alternative to traditional plasma sampling involves the collection of a drop of blood onto a filter paper card. Dried blood spot (DBS) samples have been used for over 60 years for metabolic testing in newborns [31] but recently gained popularity for use in regulated bioanalysis [32]. DBS offers the potential for self-sampling at home with a relatively painless finger stick that can provide PK data at time points unlikely to be collected (i.e., during a migraine) outside of the clinic. Additionally, DBS cards can reduce blood volume collected to ≤ 40 μL/spot, which is especially advantageous in small animal toxicokinetic studies [33]. Another advantage of DBS is that cards can usually be shipped under ambient conditions to reduce shipping costs and without need for a centrifuge or freezer at the clinic. Dried blood spots are collected through finger stick or venous blood collection followed by the application of a drop of blood onto a DBS card. Quantitation of the blood concentration is obtained by preparing standard curves/QCs on the same type of cards, punching a circle of paper within the spot, and extracting the analyte with organic solvent. Punches can be made manually with a punching tool or through the use of semi-automated punchers that cut a specific diameter circle (i.e., 1, 3, or 6 mm) and drop it into a 96 well plate for extraction. Testing to ensure analytes are stable at the condition(s) cards are stored and shipped under is an important factor to consider early in method development. Also, in addition to standard method development testing, blood pools with different hematocrits should be tested to evaluate whether there could be inherent subject-to-subject differences. A bridging study is generally required to determine the ratio between DBS and plasma-derived concentration results so comparison between matrices in possible.

5.8 Issues Specific to Endogenous Compounds

Quantitating an analyte that is inherently present in control matrix presents another bioanalytical challenge. Historically, approaches that have been used with varying degrees of success are: standard addition, quantitation against a neat standard curve, stripping the analyte of interest from the control matrix prior to use, or baseline subtraction with the detector software. An LC-MS based "surrogate analyte" approach [34] using mass spectrometry and a stable isotope labeled standard as a surrogate analyte has also been used to separate an analyte from the endogenous background levels of the analyte present in the calibration standards and QCs. In this technique, the analyst prepares a calibration curve with the stable labeled analyte and uses peak area regression to quantitate endogenous compound in the samples.

5.9 Conclusion

In this chapter we outlined the history of bioanalysis approaches, both the methodologies and instrumentation, as well as the regulations driving evolution of these approaches over the past 170 years. Critical to these techniques and the drug development process is the bioanalyst, who must understand not only the basic concepts underlying bioanalytical science, but also the regulatory requirements necessary to generate data suitable for inclusion in marketing applications for new drugs. We hope this chapter will be a helpful resource for developing this understanding. If history is any indication, we can anticipate that further evolution of the science and regulations of bioanalysis will make it an exciting discipline for years to come.

References

1. Buchanan A. Physiologic effects of the inhalation of ether. London Med Gaz. 1847;39:715–7.
2. Wagner JG. History of pharmacokinetics. Pharmacol Ther. 1981;12:537–62.
3. Folin O, Wu H. A system of blood analysis. J Biol Chem. 1919;38:81–110.

4. Haggard H. An accurate method of determining small amounts of ethyl ether in air, blood, and other fluids, together with a determination of the coefficient of distribution of ether between air and blood at various temperatures. J Biol Chem. 1923;55:131–43.
5. Shaffer P, Ronzoni E. Ether anesthesia: I. The determination of ethyl ether in air and in blood, and its distribution ratio between blood and air. J Biol Chem. 1923;57:741–60.
6. Stevenson D. Bioanalysis through the ages. Bioanalysis. 2011;3:2681–3.
7. Biollaz J, Schelling JL, Jacot Des Combes B, Brunner DB, Desponds G, Brunner HR, et al. Enalapril maleate and a lysine analogue (MK-521) in normal volunteers; relationship between plasma drug levels and the renin angiotensin system. Br J Clin Pharmacol. 1982;14:363–8.
8. US Food and Drug Administration. Guidance for industry: bioavailability and bioequivalence studies for orally administered drug products-general considerations. Available from: http://www.fda.gov/ohrms/dockets/ac/03/briefing/3995B1_07_GFI-BioAvail-BioEquiv.pdf.
9. Skelly JP. A history of biopharmaceutics in the food and Drug Administration 1968–1993. The AAPS J. 2009;12:44–50.
10. US Food and Drug Administration. Guidance for industry: process validation—general principles and practices. Available from: http://www.fda.gov/downloads/Drugs/…/Guidances/UCM070336.pdf.
11. Vuignier K, Schappler J, Veuthey J-L, Carrupt P-A, Martel S. Drug—protein binding: a critical review of analytical tools. Anal Bioanal Chem. 2010;398:53–66.
12. Wang H, Zrada M, Anderson K, Katwaru R, Harradine P, Choi B, et al. Understanding and reducing the experimental variability of in vitro plasma protein binding measurements. J Pharm Sci. 2014;103:3302–9.
13. Fung EN, Zheng N, Arnold ME, Zeng J. Effective screening approach to select esterase inhibitors used for stabilizing ester-containing prodrugs analyzed by LC-MS/MS. Bioanalysis. 2010;2:733–43.
14. Tong W, Chowdhury SK, Chen J-C, Zhong R, Alton KB, Patrick JE. Fragmentation of N-oxides (deoxygenation) in atmospheric pressure ionization: investigation of the activation process. Rapid Comm In Mass Spec. 2001;15:2085–90.
15. Holhorst M, van Amsterdam P, Heinig K, Zwanziger E, Abbott R. Stabilization of clinical samples collected for quantitative bioanalysis—a reflection from the European Bioanalysis Forum. Bioanalysis. 2015;7:333–43.
16. Kuang B, King L, Wang HF. Therapeutic monoclonal antibody concentration monitoring: free or total? Bioanalysis. 2010;2:1125–40.
17. DeSilva B, Smith W, Weiner R, Kelley M, Smolec J, Lee B, et al. Recommendation for the bioanalytical method validation of ligand-binding assays to support pharmacokinetic assessments of macromolecules. Pharm Res. 2003;20:1885–900.
18. Sennbro CJ, Knutsson M, van Amsterdam P, Timmerman P. Comparison of anticoagulant counterion impact on bioanalytical LC-MS/MS assays; results from surveys within the European Bioanalysis Forum. Bioanalysis. 2011;3:2393–9.
19. Chen Q, Tung EC, Ciccotto SL, Strauss JR, Ortiga R, Ramsay KA, Tang W. Effect of the anticoagulant ethylenediamine tetra-acetic acid (EDTA) on the estimation of pharmacokinetic parameters: a case study with tigecycline and ciprofloxacin. Xenobiotica. 2008;38:76–86.
20. European Medicines Agency. Guideline on bioanalytical method validation. Available from: http://www.ema.europa.eu/docs/en_GB/document_library/Scientific_guideline/2011/08/WC500109686.pdf.
21. US Food and Drug Administration. Guidance for industry—bioanalytical method validation draft guidance. Available from: http://www.fda.gov/downloads/drugs/guidancecomplianceregulatoryinformation/guidances/ucm368107.pdf.
22. Shou WZ, Naidong W. Post-column infusion study of the dosing vehicle effect in the liquid chromatography/tandem mass spectrometric analysis of discovery pharmacokinetic samples. Rapid Comm Mass Spec. 2003;17:589–97.
23. Ji AJ, Jiang Z, Livson Y, Davis AD, Chu JX, Weng N. Challenges in urine bioanalytical assays: overcoming nonspecific binding. Bioanalysis. 2010;2:1573–86.

24. Groff M, Riffel K, Song H, Lo M-W. Stabilization and determination of a PPAR agonist in human urine using automated 96-well liquid-liquid extraction and liquid chromatography/tandem mass spectrometry. J Chrom B. 2006;842:122–30.
25. Xu Y, Du L, Rose MJ, Fu I, Woolf EJ, Musson DG. Concerns in the development of an assay for determination of a highly conjugated adsorption-prone compound in human urine. J Chrom B. 2005;818:241–8.
26. Chen C, Bajpai L, Mollova N, Leung K. Sensitive and cost-effective LC-MS/MS method for quantitation of CVT-6883 in human urine using sodium dodecylbenzenesulfonate additive to eliminate adsorptive losses. J Chrom B. 2009;877:943–7.
27. Silvester S, Zang F. Overcoming non-specific adsorption issues for AZD9164 in human urine samples: consideration of bioanalytical and metabolite identification procedures. J Chrom B. 2012;893–894:134–43.
28. Wilson S, Kerr S, Rose M, James C. Validation of a method for the determination of AMG 579 in cerebrospinal fluid with a focus on sample collection procedures for clinical trials. J Pharm Biomed Anal. 2015;108:49–55.
29. Xue YJ, Gao H, Ji QC, Lam Z, Fang X, Lin Z, et al. Bioanalysis of drug in tissue: current status and challenges. Bioanalysis. 2012;4:2637–53.
30. Kemper CJ, Koetzner L, Koleto M. A primer for best practices in tissue preparation for bioanalysis. Bioanalysis. 2012;4:2621–36.
31. Guthrie R, Susi A. A simple phenylalanine method for detecting phenylketonuria in large populations of newborn infants. Pediatrics. 1963;32:338–43.
32. Xu Y, Woolf EJ, Agrawal NGB, Kothare P, Pucci V, Bateman KP. Merck's perspective on the implementation of dried blood spot technology in clinical drug development-why, when and how. Bioanalysis. 2013;5:341–50.
33. Beaudette P, Bateman KP. Discovery stage pharmacokinetics using dried blood spots. J Chrom B. 2004;809:153–8.
34. Li W, Cohen LH. Quantitation of endogenous analytes in biofluid without a true blank matrix. Anal Chem. 2003;75:5854–9.

Chapter 6
Fundamentals of LC-MS/MS for Regulated Bioanalysis

Gary Schultz and Jack Henion

Abstract The development of liquid chromatography-mass spectrometry technologies in the 1970s and 1980s had fundamentally changed the development of new drugs and the pharmacokinetic data quality generated during bioanalysis. A brief review of these historical developments of the interfacing of liquid chromatography with mass spectrometry transitions into the best use practices for those new to the field of bioanalysis. The chapter highlights the benefits of implementing ultra high-pressure liquid chromatography to provide improved sensitivity, selectivity, and analysis speed combined with triple quadrupole mass spectrometry. The use of appropriate solvents and solvent modifiers combined with the stationary phase appropriate for the analyte's structure is discussed and provided in an easy to read and understand manner. The introduction of the eluate to the mass spectrometer ion source is described for the two most commonly used atmospheric pressure ionization techniques in use today. The process of electrospray ionization and atmospheric pressure chemical ionization are used to create charged molecules which are directed into the vacuum region of the mass spectrometer. A step-by-step guide is provided on the optimization of the mass spectrometer settings which provides selective and sensitive selected reaction monitoring signals that can be used to quantify compounds. Best practices are discussed for system maintenance, calibration, and tuning and appropriate documentation needed to support regulatory requirements.

Keywords Liquid chromatography-mass spectrometry (LC-MS) · Pharmacokinetics (PK) · Bioanalysis · High performance liquid chromatography (HPLC) · Ultra high performance liquid chromatography (UHPLC) · Triple quadrupole mass spectrometry (QQQ) · High resolution mass spectrometry (HRMS) · Tandem mass spectrometry (MS/MS) · Atmospheric pressure ionization (API) ·

G. Schultz (✉)
Munson Technology LLC, 520 Warren Pl, Ithaca, NY 14850-3145, USA
e-mail: gary.a.schultz@gmail.com

J. Henion
R&D, Q Squared Solutions, 19 Brown Rd, Ithaca, NY 14850, USA
e-mail: henionj@advion.com

M.L. Rocci Jr. and S. Lowes (eds.), *Regulated Bioanalysis: Fundamentals and Practice*, AAPS Advances in the Pharmaceutical Sciences Series 26,
DOI 10.1007/978-3-319-54802-9_6

Electrospray ionization (ESI) · Atmospheric pressure chemical ionization (APCI) · Selected reaction monitoring (SRM)

6.1 Introduction: Interfacing Liquid Chromatography to Mass Spectrometry

High-performance liquid chromatography (HPLC) became commercially available in the early to mid-1970s. Many detectors evolved for this new separation science, but for years the UV and later, diode array detectors, became the industry standard. These detection technologies were user-friendly, such that mobile phase solvents, additives, and separation columns could be used with minimal analytical issues. Chromatographers found this new analytical technique useful for the separation of polar, nonvolatile, higher molecular weight, and thermally labile compounds; all problematic issues with gas chromatography (GC) technologies including GC/MS. The pharmaceutical industry quickly adopted HPLC for the separation and detection of drugs in the extracts of biological matrices such as urine and plasma. By the early 80s however, a practical need evolved for improved separation efficiency and selectivity [1] over that available using LC-UV. Early HPLC columns did not provide the separation efficiency of capillary GC and hence many of the HPLC chromatograms contained co-eluting components. This discrepancy illuminated a need for improved chromatographic efficiency but also, and more immediately, for a more selective detector. Because of the success mass spectrometry had demonstrated coupled with GC to provide GC/MS capability, it was tempting for some to wish for a mass spectrometer as a detector for HPLC (LC-MS). The selectivity attributes of mass spectrometry could provide an orthogonal mechanism of separation science that included molecular mass and structural information for organic molecules. But the big question then was how to do it?

HPLC practitioners resisted accepting the analytical potential of HPLC-MS (LC-MS) for several reasons. Although preparative HPLC was used for collecting relatively pure compounds for additional characterization, the quantitative determination of drugs and metabolites from biological extracts presented considerable challenges. This bioanalysis of compounds was driven in large part by pharmaceutical industry requirements for robust and reliable quantitative determination of drugs and metabolites to generate pharmacokinetic (PK) data in support of discovery, pre-clinical, and clinical investigations. Because biological extracts are chemically complex, the LC-UV chromatogram contained many co-eluting analytes. Baseline separation and quantification using chromatographic peak areas proved difficult, that is, had limited selectivity [2]. Thus method development required a chromatogram where the analyte(s) of interest and their respective chemical analog internal standards eluted in a region free of interference from endogenous components. This often required months of method development and subsequent challenging method validation.

Capillary GC/MS methods were being used concurrently, but the newly developed drugs/metabolites were increasingly difficult to measure by this method. Analysts tried chemical derivatization and other approaches to address the high polarity, thermally labile and nonvolatile nature of such compounds. Meanwhile, it was becoming increasingly clear that HPLC technologies were the preferred chromatographic approach for these more intractable compounds, but the common detectors (UV, electrochemical, fluorescence, refractive index, etc.) all lacked the more universal yet selective detection capability that a mass spectrometer detector could provide.

By the late 70s, as HPLC techniques were becoming more widely accepted, a robust and reliable commercially available LC-MS system did not exist. It would be more than 10 years with several different instrumentation and ionization improvements before a viable LC-MS system would become commercially available. This chapter provides a brief introduction to and a chronological review of these developments.

6.2 The Early LC-MS Interfaces

Early attempts at LC-MS introduced the HPLC mobile phase directly into the ion source of the mass spectrometer. Victor Tal'roze was the first to report this in the Russian Journal of Physical Chemistry, where he introduced a very small portion of the HPLC eluate into the electron ionization (EI) source of a mass spectrometer [3]. Although proof-of-principle for LC-MS had been demonstrated, the excess gas burden produced by volatilizing the HPLC mobile phase inside the EI source imposed a significant burden on the system's vacuum pumps. The next logical step came from the group of F.W. McLafferty at Cornell University, where approximately 1% of the 1 mL/min of HPLC eluate was introduced instead into a chemical ionization (CI) source [4]. This was termed direct liquid introduction (DLI) LC/MS by Hewlett-Packard which later commercialized this technique. McLafferty's experiments were accomplished using a high-resolution dual sector mass spectrometer with a 10 kV ion source. Such instruments were the purvey of only advanced research laboratories and therefore limited access to only a few analytical scientists. One of these authors (JH) envisioned interfacing LC to MS as more easily accomplished with a quadrupole mass spectrometer. By 1978, this was accomplished using a single quadrupole system with a chemical ionization source at a voltage of 5–10 V and was described as direct liquid introduction (DLI) LC-MS [5].

Beginning late in the 1970s and especially through the decade of the 80s saw significant advances in the development of LC-MS systems. EI LC-MS was described by McFadden et al. in 1976 [6], while Willoughby et al. later described the two stage momentum separator LC/MS interface (called Thermabeam by Extrel and particle beam by Hewlett-Packard) approach to EI LC-MS in 1984 [7].The so-called moving belt EI LC-MS approach was commercialized in 1979 (Finnigan) while Extrel commercialized their Thermabeam EI LC/MS system in 1984.

In 1988, Hewlett-Packard (now Agilent) commercialized their Particle Beam EI LC/MS system. It is worth noting that each of these systems was produced as single quadrupole EI mass spectrometer systems rather than the familiar atmospheric pressure ionization (API) tandem triple quadrupole mass spectrometers common today.

Early EI LC-MS was not widely adopted as a technique primarily because it limited the flow rates and mobile phase compositions of HPLC chromatography. It also challenged the vacuum pumping capabilities of EI mass spectrometers. In the late 1970s, SCIEX had developed an atmospheric pressure ionization (API) source on a triple quadrupole mass spectrometer. Originally this commercial trace atmospheric gas analyzer (TAGA) mass spectrometer system was installed in a motor home vehicle and driven around a chemical manufacturing facility to 'sniff' ambient outside air for potential toxic chemicals.

Early in the 80s, one of these authors (JH) took a Waters HPLC system along with the Hewlett-Packard DLI LC/MS probe to Peter Dawson's laboratory at the National Research Council (NRC) in Ottawa, Canada. The goal was to couple the Waters HPLC system and the DLI probe to the SCIEX TAGA 6000 API triple quadrupole mass spectrometer in the NRC laboratory to show proof-of-principle for the atmospheric pressure chemical ionization (APCI) LC-MS/MS determination of sulfa drugs in equine urine extracts 1978 [8] The experimental results demonstrating successful atmospheric pressure chemical ionization (APCI) coupled with LC-MS/MS were published in 1982, setting the stage for considerable future developments in API and tandem mass spectrometry (MS/MS) experiments.

In parallel to the practical developments and application of APCI as an effective ionization technique, another API technique was emerging from the research laboratory of Prof. John Fenn at Yale University. The potential of this electrospray ionization (ESI) LC-MS technique was revealed in 1984 with the paper by Yamashita and Fenn [9]. Initial ESI reports were not on-line LC-MS applications but involved introduction of sample solutions via infusion at flow rates less than five microliters/min—well below what chromatographers were using in the early 80s. Despite its flow rate disparity, the development was revolutionary because it provided facile ionization of previously intractable compounds as well as providing intact molecular weight information. A wider range of challenging compounds could now be determined—from thermally labile polar metabolites to peptides, proteins, industrial polymers, and many other compounds previously not amenable to either EI or CI mass spectrometry. Many improvements to electrospray rapidly followed including ion spray [10] and turbo ion spray [11]. These advancements enabled HPLC techniques to be coupled with electrospray at flow rates acceptable to chromatographers. And the rest, as they say, is history. ESI came to eclipse most prior LC-MS developments and earned Prof. John Fenn a shared Nobel Prize in 2002. With commercial development as a robust and reliable analytical technique, ESI LC-MS techniques quickly qualified for the rigors of regulated bioanalysis. Although there was a time when the regulators would not accept bioanalysis data from an LC-MS/MS method, it is now generally considered the gold standard for bioanalysis.

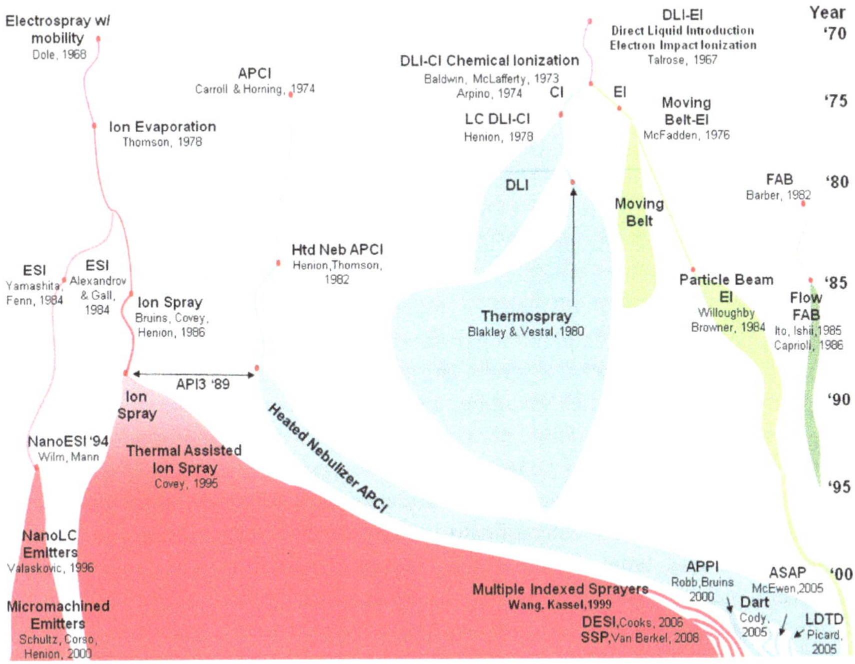

Fig. 6.1 Histomap showing a pictorial view of the early chronological developments in LC-MS techniques. Courtesy of Dr. Tom Covey, Sciex, Inc.

Figure 6.1 shows a figure by T. Covey depicting a Histomap of LC-MS developments [12]. As summarized above, the early reports by Tal'roze and McLafferty evolved to multiple alternatives and advancements which for LC-MS applications are dominated today by bioanalytical applications with ESI and APCI. There have been many other types of API sources including atmospheric pressure photoionization (APPI) [13], and desorption ionization techniques such as MALDI, but none of these have enjoyed the commercial and applications success of ESI in particular and to a lesser extent APCI LC-MS/MS.

6.3 The 3 S's of Bioanalysis—Sensitivity, Selectivity, Speed

6.3.1 Liquid Chromatography for Bioanalysis

6.3.1.1 Liquid Pumping Systems

The old adage 'Time Is Money' rings true when considering whether to purchase a HPLC or UHPLC (ultra high performance liquid chromatography) system for bioanalysis [14]. The most modern UHPLC systems can operate at fluidic pressures up to 1500 bar, which enables implementation of stationary phase particles approaching 1 μm in diameter. Smaller diameter stationary phase particles provide greater efficiency at high linear velocities than larger diameter particles, resulting in separations three (or more) times faster than separations using HPLC. This is due to the inverse relationship between particle size and efficiency and optimal flow rate. We encourage practitioners new to bioanalysis to consider UHPLC as the platform of choice for the high throughput benefits and high analytical performance characteristics discussed further below.

Advances in UHPLC efficiencies require greater emphasis on minimizing extra-column variance (i.e., plumbing dead-volumes) throughout the LC system and the interface to MS. UHPLC systems are just that; systems. They consist of integrated components of an LC, including a sample manager, an autosampler, two or more pumps, a solvent or mobile phase mixer, column oven, multi-port valves and detector(s). These integrated UHPLC systems are carefully designed to minimize extra-column variance so that the separation efficiency measured at the detector is as close to that which is theoretically achievable for the column and particle dimensions in use. Improvements in separation efficiency increase selectivity by improving the separation of endogenous components from targeted analytes. In addition to increased analysis speed and selectivity, UHPLC enhances detection sensitivity over HPLC due to the resulting improvement in signal to noise. The 3 S's of bioanalysis are achieved with UHPLC and translates directly to a lower cost of analysis.

The LC column stationary phase chemistries should not be overlooked as a key component of delivering selectivity. They are rapidly evolving to meet the needs of modern chromatography applications. The chemistries include a multitude of reversed-phase stationary phases optimized for specific compound classes. These include varying the length of the hydrophobic backbone of the stationary phase, ranging from C-4 to C-18 (for nonpolar compounds) to amino and cyanopropyl phases for medium polar to more polar compounds. Additionally, controlling analyte interaction with silica particles to further support the stationary phase influences retention order and selectivity based on hydrophilic interactions between analytes and silica solid supports. Columns that are stable at low pH (<2) and high pH (>10) can have unique applications, particularly where analyte stability is of concern. Normal-phase materials such as hydrophilic interaction liquid chromatography (HILIC) [15], nonporous particles, polymer (particles and monoliths) [16], and solid-core (nonporous particles coated with a porous layer) [17] columns are all in a continuous state of development by the column manufacturers. Each of these materials warrant the attention of the modern bioanalyst seeking ultimate performance, robustness, and analytical throughput.

All column chemistries help to achieve the selectivity needed to resolve analytes from each other and from co-eluting components. Practitioners new to LC-MS may be forgiven for thinking that chromatographic separations can be compromised because mass spectrometry offers very selective detection. While this is true, co-eluting components may interfere with ionization of the analyte of interest and cause matrix suppression or enhancement along with variability from one sample to the next. While ESI is more susceptible to these matrix effects, APCI can also suffer from it. Ultimately, the bioanalyst will often need to seek improved chromatographic separations for bioanalytical assays that are developed and validated for routine study sample analysis.

As previously noted, UHPLC analysis speed enables reduced cycle times between injections of samples. Reduced cycle times also afford faster assay development and validation. The ability to assess an assay variable through a method development batch/run in a single 8–10-hour work day rather than wait for over-night acquired data, greatly assists efficiency of the bioanalytical laboratory. Once an assay is validated, the benefits to faster sample throughput are further realized. This faster chromatography lowers the cost of ownership of the instruments for labs that are able to utilize this time savings with the analysis of more samples. This can be achieved by increasing the number of study samples within a given run, or by completing the analysis of a 200-injection run in 4 hours rather than 12–16 h (using HPLC), thereby enabling the analysis of samples for a completely different assay using the same UHPLC-MS system on the same day. Seamless software control of the LC and MS system will enable the latter and this is best achieved with the use of an LC-MS designed to work together, that is, from the same manufacturer.

6.3.2 Mobile Phase

Solvents specifically purified for use as an LC-MS solvent should be selected, though it should be noted that solvents optimal for LC-UV are not guaranteed to provide the best performance for an LC-MS analysis. Solvents developed for LC-MS undergo improved distillation and/or chemical scrubbing to remove contaminants (organic contaminants such as surfactants and plasticizers and metal ions which can form adducts to analytes) that contribute to high-baseline MS signal in the low mass range of a mass spectrum. In the earlier days of LC-MS applications, a bioanalyst could purchase what was called LC-grade solvents intended for LC with UV detectors. Clearly, if an impurity in the solvent lacks any chromophore, it would not be detected so some of these products could not be observed with UV detection, but when used as LC-MS mobile phases often showed high chemical noise. Thus, the solvent impurities not observed with UV detection could easily interfere with LC-MS analyses. Today, it is important to purchase LC-MS-grade solvents for LC-MS mobile phase applications. Solvent purity has been shown to negatively affect sensitivity, selectivity, and system robustness [18].

Mobile phase additives are used as needed to enhance chromatographic and/or mass selectivity. Volatile additives are preferred so as to reduce ion source contamination and suppression of ionization. Commonly used additives include formic acid, acetic acid, ammonium formate, ammonium acetate, and even ammonium carbonate. Acid concentrations vary from 0.02% up to a few percent. Trifluoroacetic acid (TFA) is occasionally used as an ion-pairing agent at low concentrations, alone or in combination, with a weaker acid to improve chromatographic retention of polar analytes. With the broad selection of stationary phases available, however, we advise varying the chromatographic selectivity through the selection of an alternate stationary phase to improve analyte retention rather than using TFA, which is prone to forming molecular adducts and can often reduce analyte sensitivity. When choosing HPLC mobile phase additives, the bioanalyst must consider ionization polarity of the analytes. In general, the pH of the mobile phase for positive ion ESI is neutral to acidic while for negative ion ESI mobile phase pH ranges from neutral to basic. Electrospray ionization depends upon the desired analyte ions existing in the mobile phase, so when the mobile phase contains formic acid, analytes are more likely to form positive ions. The alternative situation where ammonium hydroxide is the additive in the mobile phase tends to form negative analyte ions. Thus, the pH of the mobile phase should be considered for optimal positive versus negative ion ESI detection.

Buffer concentrations are commonly in the range of a few millimolar (mM) up to 20 mM. The buffer pH should generally be 2 pH units away (above or below) from the pK_a of the analyte(s) to ensure peak shape and improved chromatographic retention. Mobile phase preparation should be documented in a laboratory or electronic notebook, documenting date of preparation and sufficient detail of the lab ware used and reagent lot numbers should be recorded. An expiration date should be assigned to each solution prepared which reflects the expected stability for the storage conditions being used. Particular care should be taken with aqueous buffer solutions at or around physiological pH (6–8), as bacteria can grow in these solutions. Expiration dates of a week or less should be considered for neutral pH buffers. Alternatively, add an organic modifier to the buffer if possible. Bacterial growth is particularly troublesome as it can cause significant complications in the UHPLC system including plugging of frits and tubes, and premature wear on pump components.

LC systems should be purged with a buffer-free solvent prior to switching to a new buffer system to reduce the potential for particle formation due to precipitation from incompatibility between the two buffers. Additionally, mobile phase pumps should be purged with a solvent for periods when a pump system will be placed out of service for an extended period of time. Following the solvent purge, pumps, and LC systems should be stored in a solvent solution containing approximately 50% isopropanol/50% water, as isopropanol is a good wetting agent and this solution will prevent bacterial growth. A pump should never be stored without a solvent and vendor recommendations should be followed.

6.3.3 *Triple Quadrupole Mass Spectrometer*

6.3.3.1 Ion Source—Atmospheric Pressure Ionization (API)

The mass spectrometer ion source is the region wherein the eluate from an LC is introduced and analytes in the liquid phase are converted to gas-phase ions. This process is primarily conducted at atmospheric pressure using either electrospray ionization (ESI) or atmospheric pressure chemical ionization (APCI) . Electrospray ionization occurs as a solution phase process when the LC eluate is introduced into an electric field sufficient to form a charged aerosol of liquid particles. This is generally accomplished by flowing the eluate through a narrow bore stainless steel tube and applying a potential voltage of between 1000 and 5000 V. For higher LC flow rates (greater than 5 μL/min), a high linear velocity gas flows coaxially to the eluate to assist in the droplet formation process (IonSpray™) [10]. Heating the gas and the ion source facilitates droplet evaporation and the formation of gas-phase ions for molecules contained in the eluate. Figure 6.2 shows an example of such a system interfaced to a simple API inlet where the gas-phase ions are sampled from an atmospheric pressure region into an inlet leading to high vacuum and the quadrupole mass analyzer region. The mass-resolved ions exit the quadrupole where they are detected by, for example, an electron multiplier detector.

APCI is an alternative API technique where analyte ions are formed by a gas-phase process whereby analytes are first vaporized using a heated gas stream to form neutral gas-phase analyte molecules that are then ionized via a charge-transfer reaction between reagent molecules contained in the mobile phase—such as protons or, to a lesser extent, ammonium ions formed with the use of ammonium-based buffering systems. The reagent ions are formed by creating a corona discharge by applying a high voltage to a tapered metal needle within the volatilized mobile

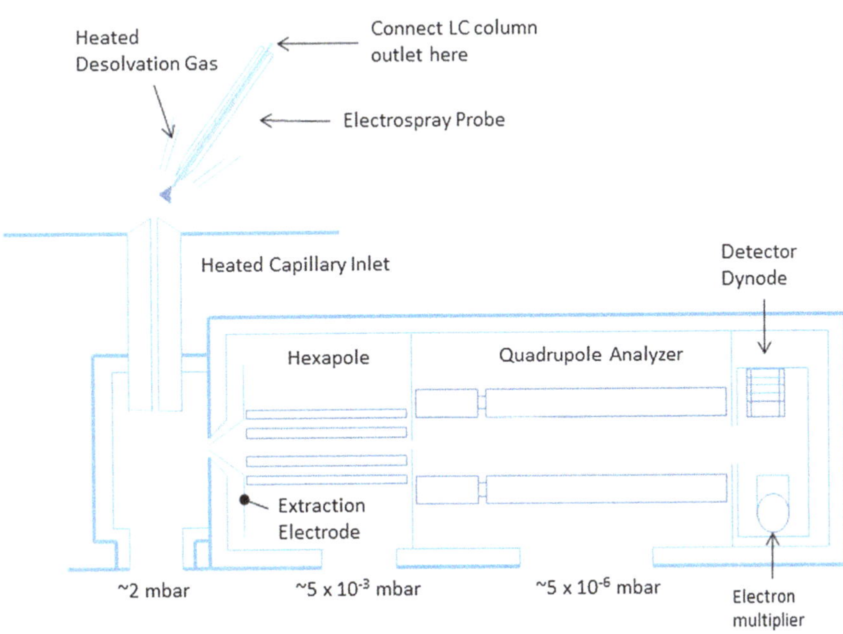

Fig. 6.2 Schematic of an atmospheric pressure chemical ionization (APCI) sprayer with the associated corona discharge needle and vacuum system inlet. Courtesy of Advion, Inc.

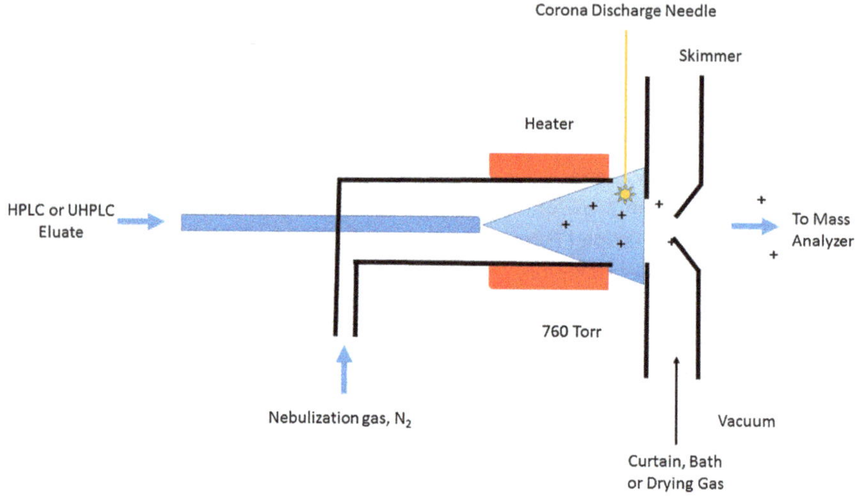

Fig. 6.3 Schematic of an atmospheric pressure chemical ionization (APCI) interface. In this example, a single quadrupole mass analyzer is shown. Courtesy of Advion, Inc.

phase (Fig. 6.3). APCI tends to work better for less polar and more volatile analytes that are less likely to form gas-phase ions with ESI. For both techniques, in general, a protonated molecule results with the application of a positive voltage to the probe or needle and a deprotonated molecule results with the application of a negative voltage. A third and less common ionization technique uses photon ionization (APPI) (vide supra) which can ionize nonpolar analytes that don't readily ionize using either ESI or APCI.

Ions formed at atmospheric pressure must then be introduced into the reduced pressure region of the mass analyzer. This is achieved primarily using a series of orifices isolating increasingly lower pressure regions using mechanical and turbo-molecular pumps to achieve the necessary pressure for optimal quadrupole operation ($\sim 5 \times 10^{-6}$ mbar). Figure 6.2 also shows ions formed at atmospheric pressure entering a vacuum region via a sample orifice followed by focusing them through a quadrupole mass analyzer to the detector.

6.4 Quadrupole Analyzer

A quadrupole is a mass-to-charge (m/z) filtering device consisting of four parallel rods upon which a radio frequency (RF) and a direct current (DC) potential voltage are applied. The ratio of the applied RF amplitude to applied DC potential determines which m/z ions entering the quadrupole are transmitted through the length of the quadrupole and exit the other end. A mass spectrum is generated by scanning the RF/DC potentials and successively transmitting each m/z value over a period of time, i.e., scan time [19]. A triple quadrupole mass analyzer consists of three quadrupole analyzers operating in series as depicted in Fig. 6.4. The first and third quadrupoles are configured to operate as scanning or transmission devices while the second, or middle, quadrupole is configured as an RF-only transmission device used as a collision cell to induce ion fragmentation with the introduction of a gas into this region of the ion optics [20].

Triple quadrupole analyzers offer many different scan types including full-scan, selected ion monitoring (SIM), product ion scan, precursor ion scan, neutral loss scan, and selected reaction monitoring (SRM). Each of these scan types is used in the field of bioanalysis at various stages during the method development process. A full-scan mass spectrum can be acquired using either the first (Q1) or third (Q3) quadrupole and can be used during an LC-MS experiment to obtain mass spectra over the entire LC separation. Each scan provides a full-scan mass spectrum of each m/z value transmitted from the ion source to the entrance of the quadrupole ion optics. Ion source and ion optics voltage settings influence which type of protonated molecules are formed and detected with the quadrupole array. These voltages can be configured to minimize the internal energy imparted to ions formed during the ionization process, also minimizing chemical bond cleavage and resulting in molecular ions. The voltage settings can be configured to impart internal energy to ions in the form of vibrational energy in the chemical bonds, resulting in bond cleavage. Remember that ions are being transmitted from atmospheric pressure in the ion source to a low pressure region of the analyzer. The droplet evaporation is

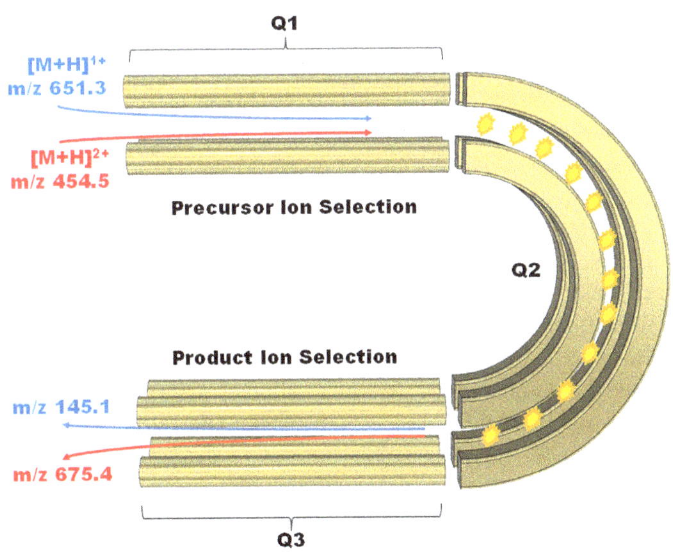

Fig. 6.4 Schematic of tandem triple quadrupole configuration typical of what is used in an LC-MS/MS bioanalytical experiment operated in the selected reaction monitoring (SRM) mode. This schematic shows two different precursor-product ion transitions which would be monitored in sequence

an endothermic process and, by itself, unlikely to induce chemical bond cleavage during droplet evaporation. However, care must be taken transmitting gas-phase ions through the ion optics of the ion source as the gas pressure is high and the energy of the collisions between solvated protonated molecules and molecules in the optics can impart sufficient energy to cause bond cleavage (fragmentation). The process is a delicate balance to minimize solvent adducts on analytes from the cooling effect while minimizing bond cleavage of the analytes.

The middle (Q2) quadrupole is most-often referred to as the collision cell of a triple quadrupole mass spectrometer. This quadrupole is configured to introduce a gas (typically highly pure argon, nitrogen or air) that will induce chemical bond cleavage of ions transmitted from Q1 into Q2 due to a combination of the pressure and kinetic energy imparted to the ions (i.e., the collision energy) for the applied potential voltage difference applied to the entrance and exit lenses of the collision cell. The collision cell pressure and collision energy in combination affect the degree of ion fragmentation and intensity of the product ions formed for each ion entering Q2. Each of these parameters is optimized during a scan to provide molecular structure data in the case of precursor and product ion scans or optimized to provide selectivity and sensitivity for selected reaction monitoring experiments. The SRM experiment is the primary tool used for quantitative bioanalysis.

6.4.1 System Calibration, Tuning, and Compound Optimization

Each quadrupole must be calibrated over a m/z range greater than that used in an analysis. This is commonly done using a calibrant solution provided by the equipment vendor. Calibrant solutions should have a traceable lot number with a certificate of analysis (CoA) identifying its composition. The calibration mass range should bracket the masses monitored in the analysis. The mass calibration procedure should also be documented in a standard operating procedure (SOP) document. A logbook should contain a record of the instrument name/number, date, solution, instrument operator, and calibration results. System calibration should be conducted at a frequency recommended by the vendor or as defined in the mass calibration SOP. For a quadrupole mass analyzer, the mass accuracy should be within 0.1 Da for each calibrant mass monitored. Mass resolution is also set and determined at the time of the mass calibration. A quadrupole mass analyzer tuned to provide unit resolution will demonstrate a peak width of between 0.5 and 0.7 Da FWHM (full-width measured at one half the height of each mass peak). Most often the optimization of resolution and calibration are automated software routines that, when completed, will provide an electronic report summary of the calibration procedure. This report should be saved and available for review as documentation the instrument is operating according to the laboratory SOP.

A mass calibration should be confirmed using a second known analyte solution as verification of a successful calibration. This is often a solution preparation of the analyte for which the mass spectrometer will be used. The analyte should be introduced into the mass spectrometer at a concentration that provides a response within the linear response range of the analyzer.

6.4.2 Mass Calibration Range and Frequency

The instrument should be calibrated from a mass-to-charge range that extends above and below the m/z values being measured. The frequency of mass calibrations should be determined based on the vendor recommendations or by established SOPs.

6.4.3 Compound Optimization

When setting up the LC-MS system for an analyte, it is common to verify the mass resolution and calibration of the instrument and to optimize the signal intensity and selectivity. This is generally done by teeing-in an analyte solution into the mobile phase held at the mobile phase composition at which the analyte elutes from the column. Parameters for tuning and optimization include adjusting the ion source temperatures and spray parameters. For most modern mass spectrometer instruments, each vendor provides suggested temperatures for each region of the ion source based on the mobile phase composition and total flow rate introduced to the ionization source. The analyte solution concentration generally is prepared between 1 and 10 μg/mL and introduced at a flow rate of 10 μL/min or less into the mobile phase at the flow rate used for the LC-SRM-MS experiment and mobile phase composition at which the analyte elutes. Dilution into the mobile phase should be sufficient to ensure the analyte response is within the linear response range of the detector.

A narrow scan range around the analyte Q1 m/z value should be configured in the software to allow for optimization of the spray parameters. Optimization of the gas flows, temperatures, and voltages applied should be conducted to achieve a stable analyte signal with variability of only a few percent. The FWHM peak width should be measured to confirm that the quadrupole calibration provides unit mass resolution for the precursor ion for each analyte in both quadrupole mass analyzers. The centroid m/z value (effectively the area weighted center of the mass-to-charge signal) for the analyte(s) should be confirmed within 0.1 Da of the exact mass for the molecular formula as a secondary confirmation of the mass calibration for each quadrupole. Selection of a product ion resulting from collision-induced dissociation of the precursor ion is conducted by optimization of the collision cell pressure and ion velocity as it passes through the high-pressure cell. The ion velocity is controlled by adjusting the collision energy, which is determined by the potential voltage difference between the entrance and exit lenses of the collision cell. Q3 should be configured to scan over a range sufficient to detect all of the product ions. For singly charged precursor ions, the Q3 quadrupole should be set to acquire data beyond the Q1 precursor so that the precursor ion is acquired in the product ion mass spectrum. We recommend acquiring and archiving an electronic copy of the confirmation scans each time the system is setup and optimized for a run. This may include a wide and narrow range Q1 and Q3 scans. Again, the Q3 mass resolution of the product ions should be measured to confirm compliance with SOP and analytical assay requirements.

Analytes with multiple protonation or deprotonation sites generally form multiply charged precursor and product ions. Distributing the analyte charge over multiple m/z ions for these molecules requires careful evaluation of the precursor and product ions selected for the method. Performing a product ion scan on the most abundant precursor ion charge-state will likely provide the best sensitivity for the analyte. Additionally, multiply charged ions often lose a charge in the collision cell to form product ions at m/z values greater than the precursor m/z value. So when acquiring product ion scans for multiply charged precursors, it is important to scan Q3 to m/z values at least two time greater than the precursor m/z value to insure that all potential product ions formed are acquired.

It is important to note that it is the combination of sensitivity and selectivity that is most important when establishing a new LC-SRM-MS method. When setting up a new method for a multiply charging analyte, the product ion mass spectrum should be acquired for (at least) the three most abundant protonated molecules observed in the Q1 scan because the highest precursor ion intensity does not always produce the most abundant product ion. Additionally, the selectivity of the selected reaction monitoring (SRM) chromatograms should be evaluated by building an acquisition method incorporating each optimized SRM and acquiring the LC-SRM-MS data on a blank sample matrix extract to evaluate the selectivity for each SRM chosen. It is helpful to note that ultimately the best method provides both high sensitivity and selectivity for the analyte(s) of interest. It is often the case that the most abundant SRM transition for a multiply charged analyte is a product ion with an m/z less than that of the precursor. However, product ions at higher m/z than the precursor m/z may provide the best selectivity. Choosing which SRM to quantify an analyte can only be assessed using matrix extract samples.

6.4.4 Data Acquisition

The number of data points acquired across each SRM transition should be sufficient to capture the transient nature of each analytes chromatogram. Statistically, 20 data points provides a 98% confidence level while 10 data points provides a 95%

confidence level. The development and use of UHPLC combined with the acquisition of multiple SRM transitions within a single LC-MS method has led to the development of improved electronics enabling the acquisition of 500 discrete SRM data points per second. Thus, a UHPLC separation with 2-s peak widths (measured at the baseline) would allow for 50 SRM experiments to be acquired simultaneously while collecting 20 data points for each SRM.

6.4.5 Chromatographic Peak Integration

Chromatographic peak integration should be performed using validated software and automated integration parameters. One universal set of integration parameters should be used for a batch of samples including standards, quality control, blank and study samples [21]. One should not conduct manual adjustment of peak integration parameters for individual samples as this can introduce operator bias into the results and is generally discouraged by regulatory authorities.

6.5 High Resolution Mass Spectrometry (HRMS)

High resolution mass spectrometry provides mass resolving powers much greater than that provided by quadruple mass spectrometry and is traditionally used for structural and molecular formula confirmation of molecules. High resolution data is used to determine the elemental composition based upon the sum of the measured accurate masses of each element and the corresponding isotope pattern for the compound. Combining high-resolution isotope pattern data with high mass accuracy can be used also to further refine the possible chemical formula for unknown molecules (such as drug metabolites) [22]. The combination of liquid chromatography with HRMS can provide the selectivity and sensitivity needed for the bioanalysis of pharmaceutical compounds.

Historically, HRMS data have been provided from magnetic sector mass spectrometers [23] as well as from Fourier Transform Ion Cyclotron (FTICR) mass spectrometers [24]. However, exciting progress has been made in recent years with the commercial availability of Quadrupole Time-of-Flight (QTOF) [25] and Orbitrap [26] mass spectrometers. These complementary technologies offer exciting opportunities for regulated bioanalysis. Although it seems certain triple quadrupoles employing SRM LC-MS will remain an important tool for regulated bioanalysis, it now appears certain that LC-HRMS techniques will also become an important bioanalysis tool both for small molecule and large molecule bioanalyses [27].

Figure 6.5 shows an example of the significant selectivity HRMS can provide for regulated bioanalysis. The protonated molecule shown at m/z 267.3 in Fig. 6.5a was acquired at unit mass resolution from a quadrupole mass spectrometer and

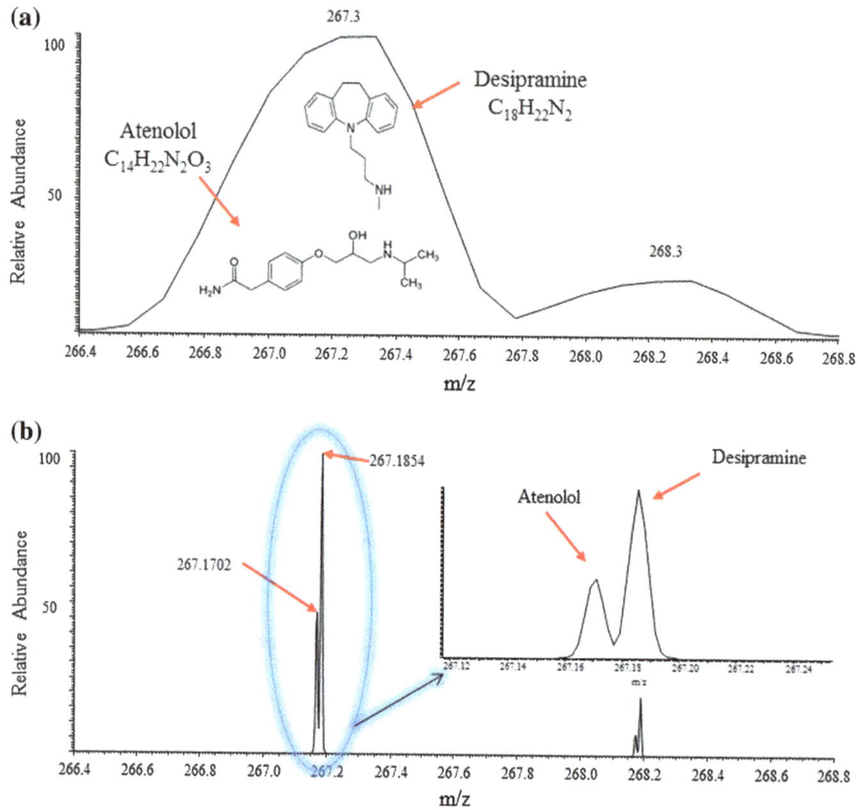

Fig. 6.5 Unit resolution on a triple quadrupole of a binary mixture of isobaric drugs (Desipramine and Atenolol) (**a**) and 35,000 resolution on an orbitrap (**b**). The inset in (**b**) shows an expanded view of the resolved isobaric compounds at 35,000 resolution. The (A + 1) ions are observed one m/z unit higher in each figure. Courtesy of Keeley Murphy and Sean Bennett, Thermo Scientific, Inc.

represents a binary mixture of isobaric drugs, desipramine, and atenolol. Both isobaric drugs are unresolved at unit resolution within this peak at m/z 267.3. When this same isobaric mixture is acquired using an orbitrap mass spectrometer using a 35,000 resolution setting, both protonated molecules of the isobaric pair are resolved with a measured mass accuracy within 3 ppm of their exact masses. The (A + 1) ions are observed one m/z unit higher in each figure. The inset in Fig. 6.5b shows an expanded view of the resolved isobaric compounds.

The tremendous analytical contributions from LC-HRMS techniques to regulated bioanalysis continue to emerge. Suffice it to say these authors believe high resolution mass spectrometry is an exciting additional tool for the future of regulated bioanalysis.

6.6 Summary and Conclusions

About 20 years ago the CEO of a major pharmaceutical company referred to LC-MS/MS as the gold standard of pharmaceutical bioanalysis. It is noteworthy that a CEO of such a large pharma company would even know about a particular analytical technique employed by his or her scientists. This fact results from the tremendous analytical horsepower of the combined techniques of HPLC (and now UHPLC) with tandem mass spectrometry employing atmospheric pressure ionization. The combined analytical capabilities of condensed-phase chromatography along with highly selective detection of targeted drugs and their metabolites by SRM LC-MS and now HRMS-LC-MS, provide unparalleled analytical capabilities compared to previous generation methodologies. As the brief overview above has described, UHPLC coupled with API tandem mass spectrometry provides a huge advance over prior regulated bioanalytical methods.

We believe SRM LC-MS bioanalytical experiments benefit from a sound knowledge base of the principles relevant to the respective technologies of UHPLC, API techniques, and mass spectrometry. Chromatographic systems must be maintained and operated according to vendor recommendations. Mobile phase solvents must be of high purity and their pedigree recorded, while chromatographic columns and plumbing connections must be properly employed. Being a good UHPLC 'plumber' can be an asset to LC-MS/MS. Similarly, experience counts when the mass spectrometer must be tuned, calibrated, and maintained to confirm all is well with the measurement system. To be sure, there is much more to conducting reliable SRM LC-MS regulated bioanalysis experiments than getting into the car and driving off!

Despite the required rigor for meeting regulatory compliant LC-MS bioanalysis, variations on the technique are now well established in supporting drug development. This technology is based upon nearly 25 years of instrumentation and technique developments and has proven to be both robust and reliable. A routine analytical technique for the pharmaceutical industry, LC-MS has now expanded widely into the clinical diagnostic laboratory and the food safety industries and even outpatient health management operations. In light of this, it seems fair to suggest that LC-MS techniques are indeed the 'gold standard' for the present as well as the future of regulated bioanalytical assays.

References

1. Vessman J. Selectivity or specificity? Validation of analytical methods from the perspective of an analytical chemist in the pharmaceutical industry. J Chromatogr. 1996;14:867–9.
2. Jörgen V, et al. Selectivity in analytical chemistry (IUPAC recommendations 2001). Pure Appl Chem. 2001;73(8):1381–6.
3. Tal'roze VL, Karpov GV. J Phys Chem. 1968; 42:1658–1664.

4. Arpino P, Baldwin MA, McLafferty FW. Liquid chromatography mass spectrometry—continuous monitoring. Biomed Mass Spectrom. 1974;1:80–2.
5. Henion JD. Drug analysis by continuously monitored liquid chromatography/mass spectrometry with a quadrupole mass spectrometer. Anal Chem. 1978;50(12):1687–93.
6. McFadden WH, Schwartz H, Evans SJ. Direct analysis of liquid chromatographic effluents. J Chromatography. 1976;122:389.
7. Willoughby RC, Browner RF. Monodisperse aerosol generation interface for combining liquid chromatography with mass spectrometry. Anal Chem. 1984;56:2626–31.
8. Henion JD, Thomson BA, Dawson PH. Determination of sulfa drugs in biological fluids by liquid chromatography/mass spectrometry/mass spectrometry. Anal Chem. 1982;54(3):451–6.
9. Yamashita M, Fenn JB. Electrospray ion source. Another variation on the free-jet theme. J Phys Chem. 1984;92:4451–9.
10. Bruins AP, Covey TR, Henion JD. Ion spray interface for combined liquid chromatography/atmospheric pressure ionization mass spectrometry. Anal Chem. 1987;59:2642–6.
11. Covey TR, Kovarik P, Jong R. The encyclopedia of mass spectrometery. In: Gross ML, Caprioli RM, editors. Molecular ionization vol. 6. Elsevier; 2007.
12. Covey T. A tribute to the scientific career of Dr. Andries Bruins. Eur J MS. 2010;16:1–10.
13. Robb DB, Blades MW. State of the art in atmospheric pressure photo ionization for LC/MS. Anal Chem Acta. 2008;627:34–49.
14. Cielecka-Piontek J, et al. UHPLC: the greening face of liquid chromatography. Chromatographia. 2013;76(21–22):1429–37.
15. Hendrickx S, Adams E, Cabooter D. Recent advances in the application of hydrophilic interaction chromatography for the analysis of biological matrices. Bioanalysis. 2015;7(22):2927–45.
16. Lubda D, et al. New developments in the application of monolithic HPLC columns. LC-GC. 2001;19(12):1186–91.
17. Venema E, et al. Packed-column hydrodynamic chromatography using 1-μm non-porous silica particles. J Chromatogr A. 1996;740(2):159–67.
18. Henion J, Brewer E, Rule G. Sample preparation for LC/MS/MS: analyzing biological and environmental samples. Anal Chem. 1998;70(19):650A–6A.
19. March R, Hugees RJ. Quadrupole storage mass spectrometery. In: Winefordner JD, editor. Chemical analysis: a series of monographs on analytical chemistry and its applications, vol. 102. New York: Wiley; 1989.
20. Boyd RK, Basic C, Bethem RA. Trace quantitative analysis by mass spectrometry. England: Wiley; 2008.
21. Zimmer D. Commentary: new US FDA draft guidance on bioanalytical method validation versus current FDA and EMA guidelines: chromatographic methods and ISR. Bioanalysis. 2014;6(1):113–9.
22. Stoll N, Schmidt E, Thurow K. Isotope pattern evaluation for the reduction of elemental compositions assigned to high-resolution mass spectral data from electrospray ionization fourier transform ion cyclotron resonance mass spectrometry. J Am Chem Soc. 2006;17(12):1692–9.
23. Henion JD, Kingston DG. Studies of substituent scrambling in diphenyl sulfide and diphenyl ether 1973, 95, 8358. J Am Chem Soc. 1973;95(8358).
24. Wang X, Emmett MR, Marshall AG. Liquid chromatography electrospray ionization fourier transform ion cyclotron resonance mass spectrometric characterization of N-linked glycans and glycopeptides. Anal Chem. 2010;82(15):6542–8.
25. Kaur S. Bioanalysis special focus issue on antibody–drug conjugates. Bioanalysis. 2013;5(9):981–3.
26. Forstenlehner IC, et al. A direct-infusion- and HPLC-ESI-Orbitrap-MS approach for the characterization of intact PEGylated proteins. Anal Chem. 2013;86(1):826–34.
27. Schultz GA, McCardle K, Neubert H. Large-scale implementation of sequential protein and peptide immunoaffinity enrichment LC/nanoLC–MS/MS for human β-nerve growth factor. Bioanalysis. 2016;8(8):753–64.

Chapter 7
Liquid Chromatography–Mass Spectrometry Assays in the Regulated Bioanalytical Laboratory

Stephanie S. Cape, Min Meng, Mohammad R. Koupaei-Abyazani and Douglas M. Fast

Abstract Data derived from robust and reproducible bioanalytical assays form the foundation of all pharmaceutical marketing approval submissions to regulatory agencies. In this chapter, bioanalytical method development, validation, and sample analysis for small molecules are discussed with emphasis on high performance liquid chromatography–mass spectrometry/mass spectrometry (LC–MS/MS) assays. Considerations for assay development include selection of analyte extraction, chromatography, and mass spectrometry parameters. Relevant aspects of validation testing and documentation are covered. Sample analysis for both good laboratory practice (animal, nonclinical) and good clinical practice (human, clinical) studies is discussed. Regulatory compliance issues are addressed for assay development, validation, and sample analysis.

Keywords High performance liquid chromatography · Mass spectrometry · Bioanalysis · LC–MS/MS · Analyte extraction · GLP · GCP · Sample analysis · Validation · Method development · Regulated bioanalysis · Documentation

S.S. Cape (✉) · M.R. Koupaei-Abyazani
Covance Laboratories, Inc., 3301 Kinsman Blvd, Madison WI 53704, USA
e-mail: stephanie.cape@covance.com

M.R. Koupaei-Abyazani
e-mail: mohammad.koupaei@covance.com

M. Meng
Covance Laboratories, Inc., 1121 East 3900 South, Salt Lake City UT 84124, USA
e-mail: min.meng@covance.com

D.M. Fast
Covance Laboratories, Inc., 109 Paradise Court, De Forest WI 53532, USA
e-mail: tdf.defo@gmail.com

M.L. Rocci Jr. and S. Lowes (eds.), *Regulated Bioanalysis: Fundamentals and Practice*, AAPS Advances in the Pharmaceutical Sciences Series 26,
DOI 10.1007/978-3-319-54802-9_7

7.1 Introduction

Previous chapters of this book introduced the reader to the regulatory, operational, and scientific aspects of bioanalysis. In this chapter, we will explore core practices of LC–MS/MS bioanalysis for small molecules, including method development, method validation, and sample analysis.

In the bioanalysis field, proper conduct of method development and validation ensures that the sample analysis procedure will produce consistently reliable results. Clear documentation of method development activities serves to explain why specific practices were invoked in bringing a proposed method to the point of validation. Regulatory auditors routinely examine the method validation data to assure that the method was validated with proper rigor and that any method limitations are well characterized. Sample analysis relies on the application of the validated method, extending demonstrated assay performance and ensuring the ability to reconstruct results with full documentation. The data generated should be accurate, repeatable, and robust. The methods developed and validated must function accurately and reproducibly not just for one study but for the extended time required for nonclinical and clinical studies, marketing application filing and consideration, drug approval, and drug post-approval studies.

Bioanalysis practices have evolved based on the first joint American Association of Pharmaceutical Scientists/US Food and Drug Administration (AAPS/FDA) workshop on bioanalytical method validation conducted in 1990 [1] and subsequent consensus meetings [2] and regulatory agency guidance documents (see Chap. 2). More recently, best global practices for validation have been recommended by an international consortium of analysts [3]. Despite these efforts for harmonization, there are still differences and nuances among laboratories. The practices outlined in this chapter present the approaches these authors have used based on their current understanding of global published documents and consensus meetings. While the criteria for acceptable validation practices are mature, they are still subject to ongoing review by regulatory authorities, including discussions about the progression of validation requirements during the drug development process [4, 5].

7.2 Method Development

This chapter provides practical guidance for a scientist developing and validating a bioanalytical method. Please refer to Chap. 6 for more detail regarding the basic concepts of MS.

7.2.1 The Scope of the Method

When developing a bioanalytical method, the first question that needs to be addressed is its intended application. If the method is to support early discovery projects, it can often be simple, fast, and generic. Typically, a discovery method consists of a protein precipitation sample preparation step followed by reversed phase liquid chromatography (LC) using a C_{18} column and a steep LC gradient program. A discovery method is considered adequate as long as the peak shape is reasonable, the carryover is manageable, and the back-calculated concentration and the coefficient of variation (CV) of the standards and quality controls (QC) meet acceptable criteria [*e.g.*, $\pm \leq 20\%$ deviation from theoretical and $\leq 20\%$ CV ($\pm \leq 25\%$ at the lower limit of quantitation (LLOQ)) respectively].

If the proposed method is intended to provide safety, efficacy, or dosage labeling data to support regulatory filing(s) (*e.g.*, good laboratory practices (GLP) or good clinical practices (GCP) assays), more comprehensive method development experiments are needed. These assays normally require enhanced targeted extraction techniques, more selective chromatography and MS conditions, and tighter acceptance criteria. Generally, per current regulatory guidance, the deviation from theoretical concentration and %CV must not deviate from specified criteria [$\pm \leq 15\%$ deviation from theoretical and $\leq 15\%$ CV ($\pm \leq 20\%$ at the LLOQ)] [6].

7.2.2 Analytes of Interest

Before developing a bioanalytical method, the target analyte(s) must be identified. In the early drug development stage, it is common to have an assay include the active drug and its metabolites. It is useful to collect all known information for these analytes such as the structure and physical-chemical properties (*e.g.*, the pKa value and log D). In addition, other critical features of the analytes, such as their propensity to adsorb or bind (nonspecific binding, NSB [7]) and their stability in neat solution and biological matrix, will be important to consider. If a new drug candidate is not stable or has poor bioavailability, it may be dosed as a prodrug. A prodrug is unstable by design so it can be metabolized into its active moiety in the body. When developing a multiple-analyte method that includes a prodrug, an active moiety and metabolite, extra caution needs to be taken. Some metabolites (*e.g.*, glucuronides and/or N-oxides) are not stable and can convert to another analyte of interest (*e.g.*, the parent drug) during sample processing or in the ion source of the mass spectrometer. If stability data are not available, a preemptive stability experiment should be conducted at the earliest possible time [8].

7.2.3 Considerations for the Biological Matrix

After the scope of the method and the analytes are determined, the next consideration is the matrix for the analysis. The most commonly used matrices are plasma, blood, serum, and urine. It is also common to develop a tissue homogenate assay from solid tissues such as lung, liver, kidney, heart, or brain. Less common matrices are cerebrospinal fluid (CSF), synovial fluid, tears, aqueous and vitreous humors, bronchoalveolar lavage fluid, feces, and other biological fluids and tissues. Plasma is the most commonly selected matrix because pharmacologic activity is primarily driven by unbound concentrations of drug in blood plasma. Blood is the alternative matrix if the analyte is heavily bound to red blood cells. Compared to plasma and serum, blood samples are more difficult to process because blood is viscous and opaque, making pipetting difficult and less accurate.

For plasma or blood samples, the anticoagulant plays an important role as well. Although different salt forms of the same anticoagulant such as Na_2EDTA versus K_2EDTA versus K_3EDTA are considered equivalent and interchangeable by regulatory standards, a change in the type of anticoagulant can influence the bioanalytical measurement. For example, a change from K_2EDTA plasma to Na-heparin plasma requires full revalidation per regulatory standards. Some less common anticoagulants such as Na-fluoride/K-oxalate can inhibit proteinases and thereby stabilize analytes.

Because urine appears clear it is possible to mistake urine as a cleaner and easier matrix to assay than plasma. Unfortunately, urine is often more difficult to handle. This also applies to other protein-deprived matrices such as CSF. Urine contains >95% water and the remaining constituents are urea, chloride, sodium, potassium, creatinine and other dissolved ions, and inorganic and organic compounds (proteins, hormones, and metabolites). Since many new drug candidates are synthetic organic compounds, they are often either not soluble in urine or have severe NSB to container walls and other incident surfaces [9].

In addition to the traditional liquid-based matrices, dried blood spot (DBS) formats, where blood samples are spotted and dried on filter paper or similar media, have recently gained attention and controversy. Because of the simplified sample handling and the potential for lower cost, DBS was introduced as a replacement for liquid plasma or blood samples. However, more research indicated that DBS has its own inherent drawbacks such as the effect of the heterogeneity of blood hematocrit [10]. Currently, consensus is to use the dried spots as an alternative matrix for specific applications when the sample volume is limited (*e.g.*, in pediatric or small rodent studies), when unique animal models are involved (*e.g.*, transgenic animals), and where nonroutine sampling is necessary (*e.g.*, at home sample collection and geographically remote studies).

7.2.4 Sample Cleanup Selection

Prior to LC–MS/MS analysis, the analyte is typically separated from major matrix components via extraction. Three common extraction techniques used in the bioanalytical laboratory are protein precipitation extraction (PPE), liquid–liquid extraction (LLE), and solid-phase extraction (SPE). PPE is the simplest extraction method in which an organic solvent such as methanol or acetonitrile is added at a minimum ratio of 1:3 (v/v) of matrix/organic solvent. This induces precipitation of the major component biomolecules while leaving the analyte(s) of interest in solution that can be separated from the precipitate. If the active drug is highly bound to protein, co-precipitation can occur. In this case, a volatile acid or base such as formic acid (FA) or ammonium hydroxide (NH_4OH) may be used to disrupt the binding, thus increasing the recovery of the drug. PPE can be conducted either in individual tubes or in 96-well plate formats. PPE in tubes is inexpensive and reliable but requires manual transfer/pipetting, vortex mixing, and centrifugation. For PPE in 96-well plates, all 96 samples can be transferred simultaneously with automated liquid handling equipment. This dramatically increases sample throughput but may add to assay costs if all 96 wells are not needed. In recent years several specialty PPE stationary phases have emerged. These plates were specifically designed to remove phospholipid interferences during PPE sample preparation. Although PPE is the most efficient and inexpensive extraction technique, it is also the most nonselective.

In contrast, LLE removes more potentially interfering endogenous matrix components. The most commonly used solvents for LLE are ethyl acetate, methyl-*tert*-butyl-ether, methylene chloride, and hexane or a combination of the above. To manipulate the polarity of the analytes, the analyst uses a volatile acid or base such as FA or NH_4OH to protonate or deprotonate the analytes. LLE can also be conducted in individual tubes or 96-well plates. For extraction in tubes, samples can be transferred either by pipetting or by the freeze-pour technique. In recent years, supported-liquid extraction (SLE) has emerged and become increasingly popular. Commercial SLE plates consist of 96 individual wells packed with a proprietary form of diatomaceous earth or another stationary phase with a high capacity for retaining aqueous samples. Because all procedural steps can be fully automated, SLE is as efficient as using PPE plates but yields much cleaner extracts. One disadvantage of SLE is the limited sample aliquot volume ($\sim$200–400 μL) because of the 96-well plate format.

SPE is another popular and selective extraction technique. It is a separation process that exploits the affinity of the analytes for a solid stationary phase. In an SPE procedure, a wash solvent is applied to remove unwanted matrix components prior to eluting the analyte of interest from the stationary phase. In general, the sorbent for an SPE stationary phase is either made of silica or a polymer backbone linked with various functional groups. Hydrocarbon chains of variable length such as C_{18}, C_8, and phenyl are commonly linked for reversed phase SPE. Quaternary ammonium or amino groups are used for anion exchange and sulfonic acid or carboxyl groups for cation exchange to allow mixed mode extraction with increased selectivity. Several well-known products with these chemistries are now commonly used in the bioanalytical laboratory.

The analyst must consider many factors when choosing a particular sample cleanup procedure for an LC–MS/MS assay [11]. If the LLOQ level is very high, it is not necessary to choose the most selective or extensive extraction procedure. A simple "dilute-and-shoot" method may serve the purpose. In general, PPE extraction should be the first choice for an assay if the target LLOQ is at or above the ng/mL concentration level and the method is used to support early drug discovery studies. If the target LLOQ is at pg/mL or sub-pg/mL concentrations or if it is used to support GLP or clinical studies, a more selective method such as LLE/SLE or SPE in a 96-well plate format should be considered. While SLE/LLE is most suitable for nonpolar analytes, SPE is the best extraction option for an assay containing a group of analytes with differing polarities such as a prodrug, an active drug, and a polar metabolite.

7.2.5 Calibrators, Quality Controls, and Internal Standards

For regulated bioanalytical methods, the precision and accuracy of the assay is evaluated by the performance of calibrators and QC samples. Regulatory guidance recommends that stocks are prepared from two independent weighings; one is used to prepare calibrators, and the second is used to prepare the QCs. A typical calibration curve contains 6–8 concentration levels to span the undiluted assay range. For the QCs, the required concentration levels are LLOQ, low (LQC, $\leq 3 \times$ LLOQ), medium [MQC, 40–60% of the upper limit of quantitation (ULOQ)], and high (HQC, 70–90% of the ULOQ). A dilution QC (DQC) may be required if samples with concentrations above the ULOQ are anticipated. The placement of the calibrators and QCs across the concentration range of the assay will be critical to proper performance of the assay in both validation and sample analysis.

For LC–MS/MS assays, the calculated concentration for QCs and samples is computed from a regression of their signal response against the response of the calibrators. The use of a suitable internal standard (ISTD) greatly assists this process by serving multiple purposes. One is to track the analyte during extraction to account for loss during handling. A second is to compensate for injection volume variability. In some cases, the ISTD may also assist in compensating for ionization suppression or enhancement during sample analysis. While this can be useful, it is best practice to remove the cause of ionization effects rather than to expect the ISTD to correct for them.

It is critical to choose appropriate ISTDs for a bioanalytical method. The ideal ISTD candidate is a stable isotope labeled (stable labeled, SIL-IS) form of the target analyte. The use of a SIL-IS is based on the assumption that the physical-chemical characteristics of a stable labeled isotope will mimic those of the unlabeled molecule except for the difference in the mass-to-charge ratios monitored in the MS.

For the selection of the stable isotope atom, a ^{13}C or ^{15}N labeled ISTD is always preferred to a deuterium labeled ISTD due to potential hydrogen-deuterium interchange. Because a SIL-IS can be expensive and time consuming to synthesize, especially for metabolites during the early phases of drug development, it may be acceptable to use a chemically similar analog of the analyte(s) as the ISTD if suitable assay performance can be demonstrated.

7.2.6 MS Tuning and Optimization

MS tuning and optimization is typically accomplished by introducing a solution of analyte (~ 500–1000 ng/mL) into the mass spectrometer by teeing it into the LC flow via a syringe pump at a 5–10 μL/min flow rate. For initial evaluation, we recommend assessing both electrospray (ESI) as well as atmospheric pressure chemical ionization (APCI) under negative or positive ionization conditions as appropriate. ESI (+) mode can be applied for the analysis of many analytes, especially those that are basic and contain primary, secondary, or tertiary amino groups. ESI (−) mode is successful for those analytes containing an acidic functional group. It is well known that the APCI interface is less prone to matrix effects because it operates under a gas phase mechanism [12]. APCI (+) or (−) should be selected preferentially over ESI if the sensitivity is sufficient for the target LLOQ and the analyte and if any metabolites are thermally stable at source operating temperatures. It is also important to carefully evaluate 1–3 predominant selective reaction monitoring (SRM) transitions throughout early method development. When selecting the precursor ion, the protonated or deprotonated ion is often selected, but ions formed by other adducts such as sodium, potassium, ammonium, or acetate may be used. When choosing the product ion, it is good practice to avoid water loss as this transition is not selective and can be easily affected by the source temperature and other factors.

7.2.7 Liquid Chromatography Development

Efficient development of chromatographic conditions requires an understanding of the chemical properties of the analytes and the characteristics of available LC columns. In general, the retention of hydrophobic compounds on alkyl-chain-based LC columns in descending order is C_{18}, C_8, C_4 for reversed phase separations. More selective columns, such as phenyl, biphenyl, or pentafluorophenylpropyl, react with any analyte aromatic ring or resonance structure via π–π interactions. Normal phase chromatography is conducted on columns using silica, cyano, or hydrophilic interaction liquid chromatography (HILIC) supports and is useful for small and polar analytes. To build a selective method, an orthogonal approach may be considered. For example, if a PPE extraction was selected for sample cleanup, a nonselective C_{18} column should be avoided if possible, particularly for large clinical studies [8].

The selection of mobile phases and modifiers is also important. Nonvolatile buffers (*e.g.*, phosphate buffers) are strongly discouraged for LC–MS/MS methods. The most common organic solvents are methanol or acetonitrile or a mixture of these two. The most commonly used acidic modifiers are FA and acetic acid. Trifluoroacetic acid (TFA) and triethylamine (TEA) are good ion pairing reagents that may be used to improve the chromatography peak shape, but they can display ion suppression and "memory effects" in the MS. It has been reported that TFA can remain in the MS for hours and has an adverse impact on subsequent assays [13]. The most common basic modifiers are NH_4OH or ammonium bicarbonate.

Although a step LC gradient from high aqueous to high organic composition can produce a sharp peak, this practice can induce carryover (see Sect. 7.2.8.2). Steep gradients can also lead to matrix effects as the analyte of interest can co-elute with interfering endogenous components. In contrast, an isocratic or a shallow gradient separation has fewer tendencies for carryover even though the peak shape may be somewhat compromised. However, the analyst should take care to ensure late eluting components do not interfere with subsequent injections.

The size of sorbent particles in the LC column has great impact on peak shape, resolution, and sensitivity. Traditional 5 μm and 3 μm particle columns are routinely used. Smaller particle size columns (*e.g.*, 1.7 μm ultra-high performance liquid chromatography (UHPLC) columns) can improve peak shape and signal-to-noise (S/N) of the analytes significantly. However, these columns have a tendency to clog if the extract is not clean and free of particulates. Leaking and blockage from less than optimal extracts jeopardize throughput and the overall performance of the assay.

7.2.8 Common Issues Encountered During Method Development

7.2.8.1 Lack of Sensitivity

Inadequate sensitivity can be addressed by improving the extraction or instrumentation. Additional sensitivity may be gained by increasing sample volume, improving extraction recovery, choosing a more selective extraction, and reducing reconstitution volume. Within the chromatographic parameters, sensitivity may be increased by reducing flowrate for an ESI assay (ESI is a concentration-dependent ionization), using smaller particle size columns (*e.g.*, UHPLC columns) to increase the S/N of the peak, or increasing organic composition by post-column addition. Multidimensional chromatography (*e.g.*, Two-Dimensional [2D] or Three-Dimensional [3D]) can significantly remove background noise and thus increase sensitivity. Within the MS parameters, multiple SRM transitions can be added together to provide greater signal or S/N than that available by using a single SRM transition.

7.2.8.2 Carryover

Carryover is residual analyte originating from a previous sample which remains detectable in subsequent analysis. Carryover issues are often the most difficult to resolve in bioanalysis, particularly when the LLOQ required is extremely low. There are two types of carryover: column carryover and injector carryover [14]. Column carryover can be resolved by using short and less retentive columns such as C_4 and C_8. An isocratic or shallow gradient can also reduce column carryover. Column backflushing via a switching valve with a strong solvent has also been reported to reduce carryover [15]. For injector carryover, a strong needle wash solution containing a mixture of strong organic solvents (*e.g.*, dimethyl sulfoxide, dimethylformamide, acetone) or an acidic or basic modifier that can degrade the analyte(s) should be utilized.

7.2.8.3 Inconsistent ISTD Response

For a bioanalytical method, the same amount of the ISTD is added into each sample aliquot. When the ISTD response is not consistent there are three possible causes: system imprecision, inconsistent extraction recovery, or variable matrix effects. To identify the cause, we recommend first performing intra-vial injections. If the intra-vial injections are very precise (*i.e.*, CV <5%) the variable ISTD response cannot be attributed to general system imprecision. Further experiments should be performed to evaluate the potential variability of extraction efficiency and/or matrix effects. The solutions for poor extraction recovery are modifying the extraction procedure, simplifying the method, and avoiding unnecessary transfer steps. If matrix effects are identified as the cause of the inconsistency, many experiments can be explored to mitigate this issue (*e.g.*, switching ionization modes, changing SRM transitions, and modifying LC conditions including mobile phases and columns). This troubleshooting strategy is clearly detailed in published case studies [16, 17].

7.2.8.4 Matrix Effects

Matrix effects are unique phenomena in bioanalytical assays that use atmospheric pressure ionization and mass spectrometer detection. Although a specific SRM transition is selected in the MS/MS experiment, many co-eluting compounds may have adverse impact on the ionization. To find out if the matrix effect is either positive (ionization enhancement) or negative (ionization suppression) an infusion experiment can be conducted [18]. In this experiment, an extracted control blank sample is injected while the analyte of interest is infused into the HPLC effluent post-column using a syringe pump. Any ion suppression or enhancement would be observed as a decrease or increase of the MS signal at the time of analyte elution. Other authors also reported that endogenous phospholipids (PL) were the primary source of plasma matrix effects when using ESI in the positive ion mode [19]. Matrix effects can be managed either via extraction or chromatography. Among the three conventional extractions, PPE is the least selective extraction technique and is therefore prone to most matrix effect. In contrast, the LLE method is more selective and can remove approximately 200 times more PLs than commonly used SPE methods [19]. Managing matrix effects via chromatography involves changing the LC column type and gradient, changing the acidic or basic modifier, or using a column switching technique to achieve two- or three-dimensional chromatography (the "heart-cut" sampling technique).

It is also best practice during method development to extract blank biological matrix aliquots from different lots and to analyze them using the proposed method to demonstrate that no substances are present at the analyte or ISTD chromatographic region that interfere with the detection of these compounds.

7.3 Validation

7.3.1 Introduction

7.3.1.1 Purpose for Validation

The purpose of bioanalytical method validation is to demonstrate that the analyte concentration measured during application of the method for sample analysis accurately reflects the concentration of the compound circulating in the body at the time of sample collection. The intended use of the concentration data helps determine the rigor and testing necessary to qualify or validate an assay that is appropriate in each context. Regulatory agencies expect most definitive nonclinical and clinical study assays to be fully validated for data submitted to support a marketing application. However, there has recently been significant discussion about an abbreviated testing approach, particularly for early nonclinical and even first-in-human studies, referred to as a fit-for-purpose or tiered approach to validation [4, 5]. This philosophy allows for flexibility and scientific judgment to be applied to ensure the method is adequately tested and will produce reliable results that can be referred to for internal decision-making or initial data submissions. As of this writing, there is no consensus among bioanalytical scientists and regulatory staff on the use of tiered validations. While tiered validation is seen as one possible solution to the increasing cost and time for drug development, the current regulatory thinking is summarized in a 2011 paper in which the author explains that "if the data generated will support regulatory action, such as assessing safety and/or efficacy, or supporting labeled-dosing instructions or patient treatment, then the data must be reliable and the analytical assays should be fully validated" [20].

7.3.1.2 Regulatory Guidances and White Papers

In this chapter, we focus on formal bioanalytical method validation as described in the guidances issued by the US FDA [6], and the European Medicines Agency (EMA) Committee for Medicinal Products for Human Use [21]. Other bioanalytical method validation regulatory documents published by authorities from countries such as Brazil (ANVISA), Japan (MHLW), Canada (TPD), and China (CFDA) are fundamentally similar to the EMA and US FDA guidance yet contain some nuances (see Chap. 2 for additional details).

These documents provide guidance for validating bioanalytical methods. The interpretation of these guidance documents is ever evolving and often refined through publication of white papers, consensus documents, and conference reports.

7.3.1.3 Regulatory Compliance

Although a formal claim of compliance to the Organisation for Economic Co-operation and Development (OECD) Principles on GLP is generally not necessary for the conduct of bioanalytical method validation, many expectations of GLP do apply. For example, actions or observations should be documented per the principles of GLP, and "all data generated during the conduct of the study should be recorded directly, promptly, accurately, and legibly by the individual entering the data. These entries should be signed or initialed and dated" [22].

During validation, the assay specifications derived from method development are rigorously tested to demonstrate that the results generated are reliable and reproducible. Additionally, analyte stability over the course and various environments of sample collection, storage, extraction, and analysis is also thoroughly interrogated during method validation.

We advise completing the method validation prior to initiating animal or human studies so that the stability of the compound(s) at sample collection and storage is well understood prior to sample collection. Therefore, planning the bioanalytical portion of the work is important to ensure adequate time to complete method development and validation prior to this milestone.

7.3.2 Required Validation Tests

7.3.2.1 Validation Requirements

Fundamental parameters that must be tested in validation include precision, accuracy, reproducibility, and selectivity of the method as well as analyte stability throughout its life cycle, from sample collection to analysis. Further tests are conducted to ensure no additional variables impact the data. See Table 7.1 for a detailed summary of validation tests. These tests and acceptance criteria can differ

Table 7.1 Detailed summary of validation tests

Validation test	Experimental requirements	Criteria
Calibration curve	• 6–8 nonzero samples spanning the quantitation range	• Simplest regression model should be applied • Measured concentration within 15% of nominal (20% at the LLOQ) • At least 75% should meet acceptance criteria for each run
Precision	• Concentration levels: LLOQ, LQC, MQC, HQC • 5 replicates per concentration level • 3 runs required (analyzed on at least 2 different days)	• Within-run CV $\leq$15% ($\leq$20% at LLOQ) • Between-run CV $\leq$15% ($\leq$20% at LLOQ)
Accuracy	• Concentration levels: LLOQ, LQC, MQC, HQC • 5 replicates per concentration level • 3 runs required (analyzed on at least 2 different days)	• Within-run mean concentration within 15% of nominal (20% at the LLOQ) • Between-run mean concentration within 15% of nominal (20% at the LLOQ) (EMA Guideline)
Run blanks	• 1 blank matrix extracted without ISTD	• No significant interference for the analyte or ISTD[a, b]
	• 1 blank matrix extracted with ISTD	• No significant interference for the analyte[a]
Selectivity	• 6 individual sources of blank matrix extracted without ISTD	• No significant interference for the analyte or ISTD[a, b]
	• 6 individual sources of blank matrix fortified to the LLOQ or low QC concentration extracted with ISTD	• Measured concentration within 15% of nominal (20% at the LLOQ)
Matrix effect	• Compare response of post-extraction spiked samples to response in absence of biological matrix • Determine matrix factor and ISTD-normalized matrix factor for 6 individual sources of blank matrix	• CV of ISTD-normalized matrix factor calculated from the 6 lots $\leq$15%
Hemolysis	• When appropriate per study design, create hemolyzed test matrix (2–3% lysed whole blood in plasma) • Concentration levels: LQC and HQC	• Evaluate matrix effect either by determining matrix factor or by assessing the ability to accurately and precisely measure the nominal concentration
Hyperlipemia	• When appropriate per study design, create, or obtain hyperlipemic plasma • Concentration levels: LQC and HQC	• Evaluate matrix effect either by determining matrix factor or by assessing the ability to accurately and precisely measure the nominal concentration
Recovery	• Determine at LQC, MQC, and HQC concentration	• Need not be 100%, but must be consistent, precise, and reproducible with acceptance defined per SOP
Dilution integrity	• Sample above ULOQ should be diluted with blank matrix • At least 5 determinations per dilution factor • Should cover the dilution applied to the study samples	• Mean accuracy within 15% of nominal and CV $\leq$15%
Run size	• Include the maximum number of samples anticipated in a single run • Concentration levels: LLOQ, LQC, MQC, HQC • 5 replicates per concentration level	• QC mean concentration within 15% of nominal (20% at the LLOQ) • QC CV $\leq$15% ($\leq$20% at LLOQ)

(continued)

Table 7.1 (continued)

Validation test	Experimental requirements	Criteria
Solution stability	• Conducted for each solvent and container material at low and high concentration to mimic handling during use for sample analysis	• Define within SOP; generally if the response of a fresh solution compares to the response of the test solution within 5–10%, this is deemed acceptable
Short-term matrix stability	• Bench-top stability: for a time period exceeding the time samples will be exposed to the condition, subject test samples to the method-defined temperature (generally room temperature or wet ice) • Concentration levels: LQC and HQC	• When measured against a freshly prepared calibration curve concentration should be within 15% of nominal
	• Freeze and thaw stability: for a minimum of 3 cycles, expose test samples to the method-specified freeze and thaw conditions • Concentration levels: LQC and HQC • Each freeze cycle should exceed 12 h	
Processed sample (extract) stability	• Compare extracted samples stored under the anticipated conditions for study samples to a freshly extracted calibration curve	• Define within SOP, generally stored extracts must meet precision and accuracy acceptance criteria
Reinjection reproducibility	• Reinject previously analyzed extracts of a calibration curve and associated precision and accuracy samples stored at method conditions	• Define within SOP, generally stored extracts must meet precision and accuracy acceptance criteria
Long-term matrix stability	• Concentration levels: LQC and HQC • Store long-term stability test samples at the method-specified temperature for a time period exceeding the time samples will be exposed to this condition	• When measured against a freshly prepared calibration curve concentration should be within 15% of nominal

[a] Significant interference (analyte): At the retention time of the analyte any response greater than or equal to 20.0% of the LLOQ response as defined by SOP. Stated in the FDA bioanalytical method validation guidance as "The analyte response at the LLOQ should be at least 5 times the response compared to blank response." [6]
[b] Significant interference (ISTD): At the retention time of the internal standard any response greater than or equal to 5.00% of the internal standard response as defined by SOP

slightly among laboratories. It is important that the validation acceptance criteria are determined and defined a priori and outlined in relevant laboratory standard operating procedures (SOPs) and/or the study-specific validation plan.

7.3.2.2 Reference Standards

Reference standards are utilized during validation to demonstrate the method's suitability for detecting and quantifying the analyte of interest within the relevant biological matrix. Test samples are prepared by spiking the reference material into blank matrix to create solutions of specific known concentrations and then aliquots of these samples are processed per the analytical method.

Subsequently, during the sample analysis phase, the analyst relies on reference standards to confirm the function of the method per run and to compare quantitatively the signal in the incurred sample against the calibrators. Therefore, it is imperative that the reference material accurately represents the compound of interest. The reference material should be structurally equivalent to the target analyte and characterized such that the identity and purity are known. The material must also be stored and used within the timeframe and per the specifications detailed on the certificate of analysis (COA).

The requirements for reference material intended for use as ISTDs are less rigorous. Because ISTD is incorporated into calibration samples, QCs, and incurred or test samples equivalently throughout the run, any change within the ISTD solution concentration has much less impact run-to-run. It is essential to demonstrate suitability for use of the ISTD and to demonstrate that the material does not interfere with measurement of the target analyte. This is often done by extracting a blank sample containing ISTD (control zero) and monitoring the analyte channel to observe any potentially interfering peaks.

7.3.2.3 Calibration Curve

The calibration curve is created by spiking known concentrations of the target analyte(s) into the same biological matrix as the intended samples and then extracting these calibration standards in the presence of ISTD. The calibration curve defines the relationship between the measured peak area ratios and the known analyte concentrations. The simplest regression equation that adequately describes this relationship should be applied. Often, this is a linear model.

The quantitative range of the assay should be chosen depending on the expected range of concentrations that will be observed in subsequent sample analysis. Due to robustness considerations and practical limitations such as carryover, the validated calibration range is usually three orders of magnitude or less.

If stability testing is being conducted, it is necessary that the calibration standards for that assay run must be prepared fresh (never frozen).

7.3.2.4 Precision and Accuracy

The precision of the method, or the closeness of replicate measurements, must be measured both within-run and between-runs. In bioanalysis, it is commonly accepted to perform a single measurement per incurred sample. Therefore, it is imperative to demonstrate the precision of the method to understand the potential error inherent in that single measurement.

Similarly, method accuracy must be demonstrated to show that the measured concentration corresponds within acceptable tolerance to the true concentration.

Precision and accuracy are evaluated by creating test sample QCs of known concentrations and measuring the ability of the method to determine these concentrations reliably and reproducibly. Five replicates per level across a minimum of four concentration levels are required. These levels should span the calibration range and include LLOQ, LQC, MQC, and HQC as determined in method development (Sect. 7.2.5). Typically, multiple aliquots of QCs are prepared in sufficient volume to span the entire validation.

Because the precision and accuracy testing during validation is intended to evaluate method performance, all valid data collected must be considered and reported.

7.3.2.5 Sensitivity

The LLOQ is the lowest standard on the calibration curve. The target LLOQ should be determined based on the needs of subsequent sample analysis. For bioequivalence studies the LLOQ should be 1/20 of C_{max} or lower [23]. During validation, precision and accuracy at the LLOQ must be evaluated as outlined in Table 7.1. Selectivity must also be confirmed through demonstration of the lack of significant interference as compared to the LLOQ response at the retention time of the analyte in individual lots of matrix.

7.3.2.6 Selectivity

Assay selectivity describes a method's ability to distinguish the target compound from potentially interfering substances. When evaluating selectivity, it is important to check multiple individual lots of matrix for interference. This testing decreases the potential that previously unforeseen interference will be present in incurred samples. Method selectivity is established by demonstrating there are no significant interfering peaks in at least six individual discrete lots of blank matrix.

In addition to demonstrating adequate selectivity from matrix components, a method for multiple-analyte assays must also demonstrate that the analytes or related metabolites do not interfere with the quantitation of each other either directly or through interconversion. For example, the analyst must use caution to ensure that prodrugs do not convert to their active moieties in vitro during sample processing. This can be tested by fortifying a sample to the ULOQ concentration, extracting per the method specifications, and evaluating the response detected in the alternative analyte channel to determine if there is significant interference present. Metabolite back conversion can be evaluated similarly if reference materials are available. In the absence of a suitable reference material for the metabolite, testing strategies utilizing incurred samples may need to be developed.

Combination, coadministered, and concomitant medications that may potentially interfere with the analytical assay should be considered. For combination and coadministered medications, when known dosage levels will be administered and reference materials are available, it is practical to conduct selectivity assessments during validation to demonstrate that the medications do not significantly interfere with the analysis of each other. The identities of drugs administered concomitantly are less defined a priori and therefore often impractical to assess during validation.

A typical approach to demonstrate that the coadministered drug does not impact the analyte chromatography or quantitation is to assess blank matrix and LQC samples fortified with the coadministered drug at the maximum expected therapeutic concentration. If the blanks are free from significant interference and the quantitation of the LQC sample is acceptable, the assay is deemed adequately selective against the coadministered drug.

7.3.2.7 Matrix Effect

Matrix effect is the potential impact that the surrounding chemical/biological environment has on the ionization and quantitation of the target analyte. The majority of validation tests are performed using a pool of biological matrix to facilitate having enough volume to prepare test samples. However, there is a risk that animal-to-animal or subject-to-subject variability can impact the assay, particularly due to different interferences or matrix constituents.

Matrix effect is evaluated by comparing the analyte peak area in the presence of biological matrix components to the analyte peak area in the absence of biological components to determine the matrix factor. This is performed for at least six individual lots of matrix. Additionally, the ISTD-normalized matrix factor is determined and must meet acceptance criteria for precision across the multiple lots of matrix. These acceptance criteria are in place to ensure that sufficiently similar peak area ratios are obtained even in the context of variable matrix components.

Other sample conditions that can cause matrix effects include hemolysis and hyperlipidemia as well as excipients or special patient populations. Effects from these conditions should also be considered, depending on the study design. For example, hyperlipidemia is generally evaluated for methods that will support human clinical trials, but this test may not be necessary to support a nonclinical study where animals are on a controlled diet. Once the appropriate test matrix is obtained or created, assessment is performed either through a quantitative approach or via a matrix effect experiment [24].

Hemolysis occurs when erythrocytes rupture, either in vivo or in vitro, and the cell contents mix with the surrounding plasma. Matrix to test the impact of hemolysis on the method performance of a plasma assay can be created by mixing a small percentage (generally 2–3%) of lysed whole blood with plasma to create a hemolyzed test matrix [25]. In cases where an impact is observed, further method development or alternative mitigation strategies are warranted [26, 27].

Hyperlipemia, the presence of excessive amounts of fat or fatty substances in the blood, can potentially impact the sample homogeneity, analyte extraction, and/or ionization efficiency. According to the European Bioanalysis Forum, there is currently no consensus within the bioanalytical community about the definition of hyperlipemic plasma [25]. Matrix to assess impact of hyperlipemia on the analytical assay performance is created by adding fat emulsion to plasma or is sourced commercially, either as postprandial plasma or by specifically requesting plasma with a concentration of triglycerides over 300 mg/dL [25].

Excipients or dosing vehicles included in the drug formulation may impact the recovery or ionization of the target analyte(s) and/or ISTDs. For example, common excipients such as polyethylene glycols cause ionization suppression [28]. These effects must be evaluated and mitigated.

The intended use of the method should be considered when designing the validation tests to be conducted. If the method will be applied to samples from a special patient population (*e.g.*, renally or hepatically impaired), care should be taken to ensure the assay is fit for this purpose.

7.3.2.8 Recovery

The sample preparation extraction recovery of the analyte(s) and ISTD from the biological matrix does not need to be 100%, but it is important that recovery is reproducible, precise, and consistent across concentration levels. Recoveries should be reasonably high to reduce the effects of random assay noise on the reported concentrations.

7.3.2.9 Dilution Integrity

A validated method must demonstrate that dilution of samples does not impact the accuracy or precision of the method. To perform this test, a sample above the ULOQ should be diluted with blank control matrix to a concentration within the quantitative range of the method.

7.3.2.10 Carryover

Carryover should be minimized during method development and monitored during all analyses to ensure the data are not impacted. In cases where carryover cannot be avoided, the samples should be placed in an order in the run that minimizes any potential impact of the carryover.

7.3.2.11 Stability

Much of the testing necessary during validation is focused on ensuring the compound is stable throughout the sample handling, processing, and analysis steps.

Solution Stability

It is critical to demonstrate the integrity of stock and intermediate solutions for the time period encompassing use. These solutions are added into the blank matrix to create the calibration standards against which the incurred sample concentrations are measured. Because the stability of the stock and intermediate solutions can impact the final results of the analysis, it is important to establish proper storage conditions and time limitations for use. The stability must be tested for each different solvent used. A bracketing approach is generally acceptable to establish stability across multiple concentrations or temperatures.

To establish solution stability, compare an aged or stressed test solution against a fresh preparation. The control and comparator solutions are diluted with an appropriate solvent to a concentration amenable for analysis. If utilizing MS as the detection mechanism, ISTD should also be included. The peak areas (or peak area ratios as appropriate if ISTD is present) should be compared between the aged and the fresh solutions. It is generally acceptable if these peak areas compare within 5–10% as defined by SOP.

Analyte Stability in Matrix

Generally, matrix stability tests at both a low and high concentration should be assessed against a freshly prepared calibration curve. The measured analyte concentration should be compared to the nominal concentration to determine acceptability.

Due to the nature of certain compounds, stabilizers or stabilization approaches must be incorporated to ensure the integrity of the compound(s) [29]. In these cases, the stabilization process must be considered in the design of the validation tests. For example, it may be necessary to modify the standard experimental design to conduct experiments on wet ice or in matrix treated with stabilizers.

Sample Collection Stability

Adequate analyte stability throughout sample collection and handling procedures is necessary to ensure the measured concentration is reflective of the circulating concentration at the time of sampling. For plasma assays, it may be necessary to demonstrate the stability of the compound in whole blood. This evaluation is typically performed by fortifying a blood sample with the analyte(s) of interest, exposing this sample to the conditions anticipated during sample collection (*e.g.*, defined time and temperature), and then processing this sample to plasma prior to extraction and analysis [30]. The response measured from this test sample is compared to a control sample. Because this testing can be complicated by partitioning of the analyte between the plasma and the erythrocytes, modified or alternate assessment strategies may be necessary [31].

Short-Term Matrix Stability

Short-term matrix stability testing includes evaluation of the integrity of the analyte (s) through multiple freeze and thaw cycles and over time at the sample processing temperature.

Samples are most commonly shipped frozen to the bioanalytical laboratory thus requiring demonstration of stability through at least one freeze and thaw cycle. Additionally, incurred sample reproducibility (ISR) testing necessitates another freeze and thaw cycle as does any additional reanalysis. Reanalysis of samples may be necessary for analytical reasons such as a failed run or a concentration initially above the ULOQ. It is therefore recommended to establish stability for at least three freeze cycles whenever possible.

Prior to extraction, samples are thawed, mixed thoroughly to ensure homogeneity, and then an aliquot is transferred to the extraction plate or tube for further processing. Stability of the analyte in matrix for the cumulative time spent at the sample processing temperature (*e.g.*, room temperature or wet ice) must be confirmed. This test is conducted by measuring the concentrations of QC samples subjected to the sample processing temperature for a time mimicking sample analysis against a freshly prepared calibration curve.

Reinjection reproducibility and processed sample stability address the reliability of results obtained upon reanalysis or after storage of sample extracts. Reinjection reproducibility is evaluated by reinjecting previously analyzed extracts to demonstrate that reanalysis does not impact the data. Processed sample stability evaluates the integrity of the analyte within the post-sample-processing matrix by measuring the concentration of stored extracts against a freshly extracted calibration curve.

Long-Term Matrix Stability

Long-term stability, sometimes referred to as frozen matrix stability, should be conducted at multiple time points to reduce the risk associated with potential instability. Especially in clinical trials, it may be necessary to store samples for long periods of time prior to analysis. In these cases, it is best to conduct some early stability time points to confirm there has been no change or trend in analyte concentration.

Ideally, long-term stability QC samples should be created prior to incurred sample collection so that frozen stability can be established either prior to or immediately following sample analysis. The long-term stability timeframe must equal or exceed the storage time of the samples prior to analysis including any reanalysis such as ISR testing.

7.3.2.12 Additional Considerations

Due to special considerations, the experiments detailed in the guidances may not be possible or practical as written. In these special circumstances, thoughtful and scientifically sound experimental design aimed at demonstrating the appropriate validity of the method within the spirit of the guidance must be applied.

Dried Spot Analysis (e.g., Dried Blood, Plasma, or Urine)

Spots of biological matrix can be collected on a specialized card and dried prior to shipping, storage, and subsequent analysis. In this case, the process for evaluating certain method performance parameters must be modified to accommodate the unique physical characteristics of the sample [32].

Microsampling

Microsampling devices typically collect 1–100 μL of liquid which can then be processed and analyzed to support pharmacokinetic (PK) studies. There are a wide variety of microsampling techniques and devices [33]. Because these techniques utilize small volumes, the impact to the animal or subject per time point is decreased. Most general validation tests are applicable to microsamples; however, certain aspects of specific devices need to be considered in design of the validation plan.

Endogenous Compounds and Biomarkers

Because an endogenous target analyte is naturally present in the matrix, modifications to the standard validation testing paradigm must be made. For example, the blank matrix may be replaced by a surrogate matrix that closely mimics the native matrix but does not contain the target analyte. Alternatively, a surrogate analyte may be employed. When performing validation of assays for endogenous compounds, special attention must be paid to the stability assessments to ensure stability is demonstrated in all relevant (*e.g.*, native and surrogate) matrices. If using both native and surrogate matrices, parallelism studies are advised to assure that analyte in both matrices reacts similarly during analysis.

The National Institutes of Health Biomarkers Definitions Working Group defines a biomarker as "a characteristic that is objectively measured and evaluated as an indicator of normal biological processes, pathogenic processes, or pharmacologic responses to a therapeutic intervention" [34]. The US FDA Bioanalytical Method Validation Guidance for Industry, released in draft form in September 2013, explains "The approach used for pharmacokinetic (PK) assays should be the starting point for validation of biomarker assays, although US FDA realizes that some characteristics may not apply or that different considerations may need to be addressed" [35].

Solid Matrices

Analysis of drugs in solid matrices such as tissues is complex due to challenges with obtaining a representative sample, efficiently extracting the analyte(s), preparing solid calibration standards and QCs with known analyte concentrations, and demonstrating stability and recovery [36]. Although the principles of bioanalytical method validation should be applied when designing the experiments to demonstrate the suitability of the assay, generally solid matrix assays are considered qualified or fit-for-purpose.

Partial Validation

Once a full validation has been conducted, a partial validation may be performed to demonstrate validity of the method after a minor change. For example, a method transfer between laboratories utilizing the same equipment is considered a partial validation. Partial validation can range from simply conducting precision and accuracy testing to nearly a full validation. When determining the degree of additional validation required due to a change (*e.g.*, a minor variation within the extraction procedure), it is important to consider if the results of each validation test would be impacted by the modification and adjust the validation plan accordingly.

Cross Validation

Cross validation is intended to demonstrate that any differences between laboratories or methodologies do not impact the data and potentially result in discrepant outcomes. When possible, cross validation should be performed by analyzing samples with known concentrations and pooled incurred samples by utilizing analytical methods available at both laboratories. The results are compared and evaluated per the validation plan.

7.3.2.13 Reporting

A formal, detailed report based on data that can be fully reconstructed is required to support the conduct of the validation. Specific requirements are detailed within regulatory instructions with nuances existing among countries.

7.4 Study Sample Analysis

7.4.1 Introduction

The first critical steps in the quantitative evaluation of compounds and their metabolites in nonclinical, biopharmaceutics, clinical pharmacology, or drug–drug interaction studies are successful development and validation of selective, sensitive, reliable, and reproducible analytical methods for the intended use [6]. The final core component, sample analysis or evaluation, provides important data on a range of endpoints (*e.g.*, assessing the PK profile of investigational medicinal products and monitoring their safety and efficacy). Consequently, it is essential that sample analysis or evaluation is performed to an acceptable standard that will ensure patient safety is not compromised and that data are reliable and accurately reported [37].

7.4.2 General Expectation for Regulated Sample Analysis

7.4.2.1 Personnel and Equipment

All staff involved in the conduct of the sample analysis and evaluation of the collected data should be appropriately educated, experienced, and trained. They should be fully aware of their role and responsibilities with respect to the task they perform as part of a nonclinical or clinical study. Staff should receive training on GLP and GCP requirements and regulations and complete periodic refresher training at least once a year. All the training should be documented and maintained within the facility [6, 37].

All instruments and equipment used to conduct sample analysis should be qualified for the intended purpose. Equipment should be regularly inspected and maintained by qualified persons. Apparatus should be cleaned, maintained, and calibrated according to the SOPs. Calibration frequency will be determined by facility management or SOP but should be designed to ensure that all equipment remains fit for purpose [37]. All routine or nonroutine maintenance, repair, and calibration and accuracy testing for facility systems or equipment and scientific instrumentation or measuring devices should be documented in logs, laboratory notebooks, work orders, and/or electronically.

7.4.2.2 Reference Standards and Reagents

The reference standard of the analyte and the ISTD used for sample analysis should be documented by a COA maintained in the study file. The criteria listed in Sect. 7.3.2.2 for method validation also apply to sample analysis. Analytical grade chemicals and reagents should be used in the laboratory. Purity, lot number, manufacturer, and expiration date are required for all stored reagents, solutions, and mixtures.

7.4.3 Sample Tracking

Samples received in a bioanalytical lab should be logged into an appropriate laboratory information management system for tracking purposes and chain of custody documentation. All shipping containers should be opened and all paperwork from the shipment including inventory and miscellaneous paperwork should be retained. Every shipment should be inspected for sample condition (frozen, thawed, or other abnormal findings observed) and sample numbers, date received, and any discrepancies should be documented. The label on each individual sample tube should have information such as protocol number, subject number, time point, and other sample demographics. For GCP compliance, individually identifiable information (e.g., subject name, initials, date of birth) should not be on the samples, and procedures to monitor for and rectify instances where that occurs should be in place.

Samples must be stored in an appropriate storage location for the temperature required according to the study protocol. The shipper and sponsor should be informed of sample shipment receipt where appropriate. The movement of samples within the laboratory during sample processing and sample analysis should be tracked using appropriate tools.

7.4.3.1 GCP Requirements

Laboratories conducting analysis of human samples for clinical trials must adhere to GCP regulations. The ICH E6 guideline defines GCP as "an international ethical and scientific quality standard for designing, conducting, recording, and reporting trials that involve human subjects. Compliance with GCP assures that the rights, safety, and well-being of trial subjects are protected and that the clinical trial data are credible" [38]. Globally, multiple regulatory authorities have published guidance documents outlining expectations for proper handling and analysis of human clinical trial samples [20, 37–41].

All staff involved in sample handling, analysis, or reporting in support of human clinical trials should be familiar with GCP requirements. Often, similar SOPs utilized for the conduct of GLP studies are applied for the analysis of human clinical trial samples with a few key additions aimed at ensuring patient safety and confidentiality. Critical documents including patient informed consent, procedure for reporting serious GCP breaches, process for blinding/unblinding study results, and escalation plans for safety concerns must be in place.

7.4.4 *Sample Analysis Procedure*

7.4.4.1 Introduction

The method development (Sect. 7.2) and validation (Sect. 7.3) should take place in advance of sample analysis. Prior to initiation of sample analysis, the sample analysis plan and analytical method must be written and approved by the responsible bioanalytical scientist. In an analysis run, samples along with calibration standards, run acceptance QCs, appropriate blanks, and system suitability samples are extracted and analyzed. The reference standard for an analyte stock solution should be weighed in duplicate and compared to ensure weighing and dissolution precision. Stock solutions comparison is usually performed by using a corresponding method (*e.g.*, UV, LC–MS) and the results should be within 5% of each other to be acceptable. Calibration standards and QC samples should be prepared from separate stock solutions. Study samples in an analysis run should be bracketed by acceptable standard or QC samples.

7.4.4.2 Calibration Standards

Calibration standards are prepared in the same matrix as the intended study samples and a calibration curve should consist of a minimum of six nonzero acceptable standard points covering the entire assay range. We recommend that a calibration curve should be comprised of two standard points below the LQC sample and two standard points above the HQC sample. Standards can be injected in singlet or duplicate per the SOP used for the study conduct if the choice is validated. One should not extrapolate the curve above the highest calibration standard or below the lowest calibration standard.

7.4.4.3 Run Acceptance QCs

Run acceptance QCs are intended to monitor continuously the performance of the method to ensure it is working satisfactorily throughout the course of the analysis. QCs are used as indicators of a properly prepared standard curve that is sufficiently sensitive and accurate to measure the actual concentrations of the analyte in the study samples. The number and concentrations of QC samples in a run depend on the total number of samples and where their anticipated concentrations fall within the curve range. We recommend preparing QCs at a minimum of three concentrations (LQC, MQC, HQC) that should be as validated in the method (Sects. 7.2.5 and 7.3.2.4). DQC samples diluted with a factor validated at or above the highest dilution factor used for study samples should also be included in a run where study samples with concentrations above the ULOQ are assayed.

For an analysis where the study data fall within a small portion of the calibration curve, it is possible that none of the regular QC sample concentrations (LQC, MQC, HQC) are close to the concentrations of the study samples. This results in limiting the monitoring power of the QC samples. If the sample concentrations are known or anticipated to be within a narrow segment of the validated curve range, additional QC samples should be added to the run to monitor the performance of the assay near the actual concentrations of the study samples. Any additional QC levels should be validated in at least one precision and accuracy run. An alternative approach is to revise the calibration curve range. The new curve range must be validated but a full validation is not usually required if the new range is within the previously validated curve concentrations. We recommended strongly that this latter approach be followed for bioequivalence and bioavailability studies where study sample concentrations can be anticipated.

7.4.4.4 System Suitability Samples

A system suitability test evaluates the performance of the analytical system prior to use. This test is performed based on the specifications defined in applicable SOPs and the analytical method. The test is conducted at a minimum by injecting processed samples at the LLOQ, ULOQ, and a blank sample after the ULOQ sample. These samples must not be study samples from the current or future runs. Instead of study samples, separate samples should be prepared for this test. If study samples must be used from previous analysis runs, samples should be pooled to eliminate the possibility of generating additional sample concentration data for a specific study sample. The sensitivity of the system is evaluated by the S/N at the LLOQ. The chromatographic performance (peak shape and retention time) is assessed by monitoring both LLOQ and ULOQ injections. A processed matrix blank sample injected after the ULOQ is intended to assess carryover prior to starting the sample analysis run. If any of the parameters (S/N, peak shape, retention time, and carryover) is not deemed appropriate or does not meet the requirements outlined in the method or SOP, remedial action may be required and such actions should be addressed in a laboratory SOP and properly documented.

7.4.4.5 Assay Run Acceptance Criteria

The criteria outlined below are required for a sample analysis run to be considered acceptable. If a run does not meet the acceptance criteria, the reason for run rejection must be documented in the raw data and the run (calibration standards, QCs, and samples) should be prepared again and reanalyzed. In runs with multiple analytes, the acceptability of the results for each analyte is assessed independently. Failure of one or more of the multiple analytes to meet acceptance criteria does not invalidate data for the other passing analytes.

7.4.4.6 Calibration Standards

The back-calculated values of calibration standards should be within ±15.0% of theoretical concentrations. The LLOQ standard is acceptable if it is within ±20.0%. At least 75% of the standards must meet acceptance criteria. A standard may be excluded from the curve if the analyst has recorded a problem with sample preparation, documented an instrument malfunction or chromatographic issue, or the back-calculated concentration is not within the acceptable criteria described above. A minimum of six nonzero standard points should remain in a calibration curve for the run to be considered acceptable. The LLOQ or ULOQ may be deleted and a revised curve range used if all the QC samples are within the revised curve range and at least six standards remain in the curve. The reason for exclusion of any standard curve point must be documented. An entire analytical run is prepared again and reanalyzed when the calibration curve does not meet the acceptance criteria.

7.4.4.7 Run Acceptance QCs

The minimum number of run acceptance QC samples in each run should be either at least 5% of the number of unknown samples or six (duplicates of LQC, MQC and HQC), whichever is greater. The QC samples should be dispersed throughout the run. At least four of six QC samples and one-half of the QC samples at each concentration should be within ±15.0% of the theoretical concentrations for the run to be accepted. If more than two QC samples at each concentration are used, at least 2/3 of the replicates at each concentration should be within the acceptable range. In cases where diluted samples are analyzed in a run, at least half of the diluted QC samples must be within ±15.0% of theoretical value to accept the diluted sample data. The entire analytical run is prepared again and reanalyzed when the run acceptance QC samples do not meet the acceptance criteria. If only the DQC samples fail, however, the run may pass but any diluted sample should be reanalyzed.

7.4.4.8 Blank Samples to Monitor Interferences, Carryover, and Contamination

To demonstrate that no substances are present that interfere with the detection of the analyte(s) or ISTD(s), a monitoring system should be incorporated into each analytical run. Samples of blank biological matrix used to prepare the standard curve should be analyzed along with the study samples according to the method in each run. A blank sample should be spiked with ISTD (control zero) and analyzed to demonstrate that the ISTD does not interfere with the detection of the analyte. Matrix blanks included in the run also assist in monitoring any contamination or carryover during the sample processing. A blank sample immediately after the highest calibrator is used to assess carryover. A reagent blank is also processed to ensure a lack of interference in reagents used for sample preparation and analysis.

Blank matrix, control zero, and reagent blanks in a run must have no significant interference (any peak present must be $\leq 20.0\%$ of the LLOQ response) at the retention time of the analyte of interest. Additionally, any interference at the ISTD retention time in blank samples must be $\leq 5.0\%$ of the ISTD response. If the above criteria are not met, the run should be rejected or addressed according to applicable SOPs.

In cases where an assay exhibits carryover that cannot be resolved by changes to extraction or chromatography, it may be possible to reduce the impact by sample ordering (typically by low to high expected concentrations) or interspersing additional blank samples between study samples.

7.4.5 *Sample Reanalysis*

There are a number of reasons for which individual samples or the entire run should be reanalyzed in order to obtain a valid result to report for each sample. The repeat analysis may be due to technical failure or due to anomalous results. Any reanalysis must be thoroughly documented with appropriate scientific justification. Methods that require frequent reanalysis of the entire run should be assessed to assure appropriate assay performance and may require redevelopment and revalidation. Two common reasons for reanalysis are listed here.

7.4.5.1 Reanalysis for Technical Reasons

- The entire analysis run should be repeated if the run does not meet the acceptance criteria (failed calibration curve, failed run acceptance QC samples, extensive carryover, or contamination)
- Sample concentration falls outside of the calibration range
- Samples that are not bracketed by acceptable standard or QC samples
- Poor chromatography
- Instrument malfunction
- Sample processing errors.

7.4.5.2 Reanalysis for Anomalous Results

It is best practice to limit the number of samples reanalyzed for PK reasons to avoid the appearance of result selection. In bioequivalence studies, any reanalysis of study samples should be avoided unless clear technical reasons are observed and documented so there can be no perception of matching study sample results to expected results. However, because of the possibility of sample switching during handling or errors in dosing or sample collection, it is good scientific practice to have an SOP in place defining how to deal with clearly anomalous assay results such as:

- An unexpected quantifiable concentration in a sample where no analyte concentration is anticipated (*e.g.*, analyte concentration in samples from a control group or in pre-dose samples)
- An expected quantifiable concentration not present in samples where significant analyte concentration is expected.

7.4.5.3 Interference Check for Coadministered Drug

In clinical studies where two or more drugs are coadministered, an interference check should be conducted to determine if a coadministered drug interferes with the quantitation of the analyte of interest. These tests may be conducted during method validation and are described in Sect. 7.3.2.6. If needed, testing should be performed prior to initiation of sample analysis for every analyte in a validated assay used for sample analysis of clinical samples.

7.4.6 Incurred Sample Reproducibility (ISR)

The purpose of this test is to demonstrate the reproducibility of an assay used for sample analysis in nonclinical and clinical studies based on the fact that QC samples do not always adequately mimic study samples. This topic was first discussed at the Crystal City Conference in May 2006, attended by bioanalytical experts from the US FDA, pharmaceutical companies, academic representatives, and contract research organizations because of observed discrepancies between original and repeat analysis of study samples even though standards and QC samples were within acceptance criteria. Following this event, several meetings and workshops were organized to harmonize activities around ISR testing.

7.4.6.1 ISR Procedure

In general, the ISR test is performed for an assay used to support a GLP study and other regulated clinical and nonclinical studies. For nonclinical studies, the ISR evaluation is usually carried out once per species, per method, per matrix, per laboratory and preferably in the first definitive toxicokinetic (TK) study. In clinical studies, we recommend performing this test for all bioequivalence studies and each study type such as healthy volunteer, patient population, and drug–drug interaction studies.

7.4.6.2 Selection and Number of ISR Samples

ISR samples should be selected from incurred samples for which valid results have already been obtained. We recommend selecting two samples from each subject at the following concentrations—one preferably near C_{max} and the other equal to or greater than the LQC concentration. Samples with sufficient volume should be selected to allow the test to be performed in a similar fashion as the initial test (full volume or by dilution). The selected ISR samples must be within the established stability.

In general, for studies with less than 1000 samples, a minimum of 10% of the study samples should be reanalyzed. For studies with more than 1000 samples, a minimum of 100 samples plus an additional 5% of the samples exceeding 1000 is recommended for ISR testing.

The ISR test should be performed as early as possible in studies with a large number of samples and on a regular basis for studies conducted over a long period of time to ensure method reproducibility is sustained. For smaller studies the test may be carried out at the end of the study.

7.4.6.3 Acceptance Criteria

Two-thirds of the repeat samples should have a difference within 20% where the difference between the repeat and original result is calculated as follows [42]:

$$\text{Difference }(\%) = \left[\frac{[\text{Repeat Result} - \text{Original Result}]}{[(\text{Original Result} + \text{Repeat Result}) \div 2]}\right] \times 100$$

The acceptance criteria are applied to the overall ISR results and not to individual ISR runs. Additionally, the ISR assessment does not determine study acceptance or rejection; however, if the test fails, an investigation must be conducted and documented.

7.4.6.4 ISR Investigation

The initial step during an ISR investigation is to compare the repeat and original runs to determine if any change in conditions has contributed to the failure of the ISR test. These include, but are not limited to, differences in reagents, matrix lots, consumables, and stock solutions used or instability or conversion of the analyte. Other possible causes of failure may involve analyst error and/or contamination of samples, reference materials, or solutions and reagents. Additional factors that may contribute to failure of an ISR test are poor solubility, incomplete thawing, and/or insufficient mixing of samples. As part of the investigation, the ISR run may be repeated after careful step-by-step review of the method and implementation of additional precautions. This experiment provides a basis for comparison of the repeat and original results with those of the investigation run. The objective of the ISR investigation is to provide sufficient confidence in the sample concentrations being reported for a study. The internal project team or study sponsor should be consulted prior to designing any experiment for ISR investigation. Only the original results should be reported for PK/TK calculations. The repeat and any investigation results are used for comparison purposes and should not be used as the final reported data.

7.4.7 Documentation and Reporting

7.4.7.1 Documentation

It is essential to keep records of all aspects of bioanalytical sample analysis as part of a nonclinical, biopharmaceutics, clinical pharmacology, or drug–drug interaction study. The data generated in each step of a sample analysis study should be properly documented in detail and available to allow reconstruction of the study. This includes, but is not limited to, a description of the reference standard materials, blank matrix biological materials used to prepare calibration standards and QC samples, reagents, solutions, apparatus, preparation of solutions, calibration standard and QC samples, sample storage, sample tracking, sample processing, and analytical conditions.

All raw data, documentation, records, outlines, and the original signed copy of the final bioanalytical report should be stored per the test facility SOPs.

7.4.7.2 Reporting

At the conclusion of the bioanalysis and any required repeat analyses, a final study report should be prepared. A typical sample analysis report should contain the following:

- Study title
- Study number
- Bioanalytical responsible scientist
- Test facility where bioanalysis was conducted
- Sponsor of the study
- Abbreviations used
- A summary of the work conducted and the results found
- Objective of the study
- Materials and methods
- Results (data tables)
- Discussion
- Sample chromatograms

- Retention of records and specimens
- Protocol deviations
- References (if applicable)
- All appropriate amendments and appendices
- Conclusion.

The section of the report containing the study results should include the following information:

- A summary table of all runs
- Calibration curve parameters
- Back-calculated concentrations and overall accuracy and precision of calibration standards and QC samples
- Final sample concentration results
- Reassay results.

References

1. Shah VP, Midha KK, Dighe SV, et al. Analytical methods validation: bioavailability, bioequivalence and pharmacokinetic studies. Pharm Res. 1992;9:588–92.
2. Shah VP. The history of bioanalytical method validation and regulation: evolution of a guidance document on bioanalytical methods validation. AAPS J. 2007;9(1):E43–7.
3. Timmerman P, Arnold M, DeSilva B, Garofolo F, Golob M, van Amsterdam P, et al. Introduction to the proposals from the Global Bioanalysis Consortium Harmonization Team. AAPS J. 2014;16:1159–61. doi:10.1208/s12248-014-9609-4.
4. Timmerman P, White S, McDougall S, Kall MA, Smeraglia J, Fjording MS, et al. Tiered approach into practice: scientific validation for chromatography-based assays in early development-a recommendation from the European Bioanalysis Forum. Bioanalysis. 2015;7:2387–98.
5. Lowes S, Hucker R, Jemal M, Marini JC, Rezende VM, Shoup R, et al. Tiered approaches to chromatographic bioanalytical method performance evaluation: recommendation for best practices and harmonization from the Global Bioanalysis Consortium Harmonization Team. AAPS J. 2015;17:17–23. doi:10.1208/s12248-014-9656-x.
6. US Department of Health and Human Services, Food and Drug Administration, Center for Drug Evaluation and Research (CDER), Center for Veterinary Medicine (CVM), Guidance for industry, bioanalytical method validation. Rockville, MD, USA. May 2001.
7. Ji AJ, Jiang Z, Livson Y, Davis JA, Chu JX, Weng N. Challenges in urine bioanalytical assays: overcoming nonspecific binding. Bioanalysis. 2010;2:1573–86.
8. Meng M, Wang L, Voelker T, Reuschel S, Van Horne KC, Bennett PK. A systematic approach for developing a robust LC-MS/MS method for bioanalysis. Bioanalysis. 2013;5:91–115.
9. Rodila R, Kim GE, Fan L, Chang MS, Zhang J, Wu H, et al. HPLC-MS/MS determination of a hardly soluble drug in human urine through drug-albumin binding assisted dissolution. J Chromatogr B. 2008;872:128–32.
10. Suva MA. A brief review on dried blood spots applications in drug development. J Pharm BioSci. 2014;1:17–23.
11. Bylda C, Thiele R, Kobold U, Volmer DA. Recent advances in sample preparation techniques to overcome difficulties encountered during quantitative analysis of small molecules from biofluids using LC-MS/MS. Analyst. 2014;139:2265–76.

12. Mei H, Hsieh Y, Nardo C, Xu X, Wang S, Ng K, Korfmacher WA. Investigation of matrix effects in bioanalytical high-performance liquid chromatography/tandem mass spectrometric assays: application to drug discovery. Rapid Commun Mass Spectrom. 2003;17:97–103. doi:10.1002/rcm.876.
13. Shou WZ, Weng N. Simple means to alleviate sensitivity loss by trifluoroacetic acid (TFA) mobile phases in the hydrophilic interaction chromatography-electrospray tandem mass spectrometric (HILIC-ESI/MS/MS) bioanalysis of basic compounds. J Chromatogr B. 2005;825:186–92.
14. Hughes NC, Wong EY, Fan J, Bajaj N. Determination of carryover and contamination for mass spectrometry-based chromatographic assays. AAPS J. 2007;9:E353–60.
15. Chen S, Zhang J, Liao D, Qian MG. A novel dynamic flush method to reduce column-related carryover. J Chromatogr Sci. 2014;52:950–3. doi:10.1093/chromsci/bmt111.
16. Meng M, Tan L, Volker T, Adler R, Carter S, Bennett PK. Overcome non-phospholipid related matrix effect and continuing improvement for a LC/MS/MS assay for the quantification of naltrexone and 6 β-naltrexol in human plasma. Seattle, Washington: AAPS Annual Meeting and Exposition. 2009; Poster T2095.
17. Tan A, Hussain S, Musuku A, Massé R. Internal standard response variations during incurred sample analysis by LC-MS/MS: case by case trouble shooting. J Chromatogr B. 2009;877:3201–9.
18. King R, Bonfiglio R, Fernandez-Metzler C, Miller-Stein C, Olah T. Mechanistic investigation of ionization suppression in electrospray ionization. J Am Soc Mass Spectrom. 2000;11:942–50.
19. Bennett PK, Van Horne KC. Identification of the major endogenous and persistent compounds in plasma, serum, and tissue that cause matrix effects with electrospray LC/MS techniques. Salt Lake City, Utah: AAPS Annual Meeting and Exposition. 2003; Poster R6006.
20. Booth B. When do you need a validated assay? Bioanalysis. 2011;3:2729–30. doi:10.4155/bio.11.250.
21. European Medicines Agency, Committee for Medicinal Products for Human Use. Guideline on bioanalytical method validation. EMEA/CHMP/EWP/192217/2009. London, UK. 2011.
22. Organisation for Economic Cooperation and Development. Environment Directorate. OECD series on principles of good laboratory practice and compliance monitoring. OECD principles on good laboratory practice. ENV/MC/CHEM(98) 17. Paris, France. 1998.
23. European Medicines Agency, Committee For Medicinal Products for Human Use. Guideline on the investigation of bioequivalence. CPMP/EWP/QWP/1401/98 Rev. 1/ Corr **. London, UK. 2010.
24. Meng M, Carter S, Bennett PK. LC-MS bioanalysis of drugs in hemolyzed and lipemic samples. In: Li W, Zhang J, Tse FLS, editors. Handbook of LC-MS bioanalysis. Best practices, experimental protocols and regulation. Hoboken, NJ: Wiley. 2013. pp. 369–377.
25. Ingelse B, Barroso B, Gray N, Jakob-Rodamer V, Kingsley C, Sykora C, et al. European Bioanalysis Forum: recommendation on dealing with hemolyzed and hyperlipidemic matrices. Bioanalysis. 2014;6:3113–20. doi:10.4155/bio.14.252.
26. Tang D, Thomas E. Strategies for dealing with hemolyzed samples in regulated LC-MS/MS bioanalysis. Bioanalysis. 2012;4:2715–24. doi:10.4155/bio.12.229.
27. Bérubé ER, Taillon MP, Furtado M, Garofolo F. Impact of sample hemolysis on drug stability in regulated bioanalysis. Bioanalysis. 2011;3:2097–105. doi:10.4155/bio.11.190.
28. Li F, Ewles M, Pelzer M, Brus T, Ledvina A, Gray N, et al. Case studies: the impact of nonanalyte components on LC-MS/MS-based bioanalysis: strategies for identifying and overcoming matrix effects. Bioanalysis. 2013;5:2409–41. doi:10.4155/bio.13.201.
29. Li W, Zhang J, Tse FL. Strategies in quantitative LC-MS/MS analysis of unstable small molecules in biological matrices. Biomed Chromatogr. 2011;25:258–77. doi:10.1002/bmc.1572.

30. Lowes S, Jersey J, Shoup R, Garofolo F, Savoie N, Mortz E, et al. Recommendations on: internal standard criteria, stability, incurred sample reanalysis and recent 483s by the Global CRO Council for Bioanalysis. Bioanalysis. 2011;3:1323–32. doi:10.4155/bio.11.135.
31. Freisleben A, Brudny-Klöppel M, Mulder H, de Vries R, de Zwart M, Timmerman P. Blood stability testing: European Bioanalysis Forum view on current challenges for regulated bioanalysis. Bioanalysis. 2011;3:1333–6. doi:10.4155/bio.11.121.
32. Timmerman P, White S, Globig S, Lüdtke S, Brunet L, Smeraglia J. EBF recommendation on the validation of bioanalytical methods for dried blood spots. Bioanalysis. 2011;3:1567–75. doi:10.4155/bio.11.132.
33. Musteata FM. Pharmacokinetic applications of microdevices and microsampling techniques. Bioanalysis. 2009;1:171–85. doi:10.4155/bio.09.18.
34. Biomarkers Definitions Working Group. Biomarkers and surrogate endpoints: preferred definitions and conceptual framework. Clin Pharmacol Ther. 2001;69:89–95.
35. US Department of Health and Human Services, US FDA, Center for Drug Evaluation and Research, Center for Veterinary Medicine. Draft guidance: Guidance for industry, bioanalytical method validation, Rockville, MD, USA. 2013. http://www.fda.gov/downloads/Drugs/GuidanceComplianceRegulatoryInformation/Guidances/UCM368107.pdf.
36. Xue YJ, Gao H, Ji QC, Lam Z, Fang X, Lin ZJ, et al. Bioanalysis of drug in tissue: current status and challenges. Bioanalysis. 2012;4:2637–53. doi:10.4155/bio.12.252.
37. European Medicines Agency, GCP Inspectors Working Group. Reflection paper for laboratories that perform the analysis or evaluation of clinical trial samples. London, UK. 2012.
38. International Council for Harmonisation of Technical Requirements for Pharmaceuticals for Human Use (ICH). Guideline for good clinical practice E6(R1). Geneva, Switzerland. 1996.
39. Medicines and Healthcare products Regulatory Agency (MHRA). Guidance on the maintenance of regulatory compliance in laboratories that perform the analysis or evaluation of clinical trial samples. Issue 1. London, UK. 2009. http://webarchive.nationalarchives.gov.uk/20141205150130/, http://www.mhra.gov.uk/Howweregulate/Medicines/Inspectionandstandards/GoodClinicalPracticeforClinicalLaboratories/index.htm.
40. Medicines and Healthcare products Regulatory Agency (MHRA). Guidance for the notification of serious breaches of GCP or the trial protocol. London, UK. 2014. https://www.gov.uk/government/uploads/system/uploads/attachment_data/file/404588/GCP_serious_breaches_guide.pdf.
41. Medicines and Healthcare products Regulatory Agency (MHRA). Guidance document on archiving. London, UK. 2006. http://webarchive.nationalarchives.gov.uk/20080814090145/, http://mhra.gov.uk/home/idcplg?IdcService=GET_FILE&dDocName=CON2024521&RevisionSelectionMethod=Latest.
42. Fast DM, Kelley M, Viswanathan CT, O'Shaughnessy J, King SP, Chaudhary A, et al. Workshop report and follow-up—AAPS Workshop on current topics in GLP Bioanalysis: assay reproducibility for incurred samples—implications of Crystal City recommendations. AAPS J. 2009;11:238–41. doi:10.1208/s12248-009-9100-9.

Chapter 8
Development and Maintenance of Critical Reagents for Ligand Binding Assays to Support Regulatory-Compliant Bioanalysis

William Nowatzke, Franklin Spriggs, Van Fitzgerald, Aleks Davis and Ronald R. Bowsher

Abstract The FDA approved the first biotherapeutic (synthetic insulin) in 1982. Thirty-five years later, this novel class of pharmaceuticals has improved the quality of life for millions of individuals. While drug discovery of biotherapeutics and biosimilars was originally dominated by small biotechnology companies, today nearly every major pharmaceutical company in the world is engaged in this effort. Biotherapeutic programs present unique challenges to drug development, including generating supportive data to demonstrate safety and efficacy in the target population. Bioanalysis of traditional chemical pharmaceutics may be achieved after organic extraction from biological fluids, thus removing potentially interfering substances. Furthermore, strategic decisions based on the structure of the molecule often facilitates the development of an analytical method utilizing sensitive mechanical instrumentation. Biotherapeutics are often composed of amino acids whose functionality depends on complex structure. These compounds are measured via protein-protein interactions requiring unique reagents for each drug program.

W. Nowatzke (✉) · V. Fitzgerald
Radix BioSolutions, 111 Cooperative Way, Suite 120, Georgetown, TX 78626, USA
e-mail: bill.nowatzke@radixbiosolutions.com

V. Fitzgerald
e-mail: vfitz@radixbiosolutions.com

F. Spriggs
Ligand Binding Assays, AIT Bioscience, 7840 Innovation Boulevard, Indianapolis, IN 46278, USA
e-mail: fspriggs@aitbioscience.com

A. Davis · R.R. Bowsher
B2S Life Sciences, 97 East Monroe Street, Franklin, IN 46131, USA
e-mail: Aleks.Davis@B2SLifeSciences.com

R.R. Bowsher
e-mail: Ron.Bowsher@B2SLifeSciences.com

M.L. Rocci Jr. and S. Lowes (eds.), *Regulated Bioanalysis: Fundamentals and Practice*, AAPS Advances in the Pharmaceutical Sciences Series 26,
DOI 10.1007/978-3-319-54802-9_8

In this chapter, we describe these interactions and focus on the importance of generating and maintaining high-quality reagents, a process known as reagent lifecycle management.

Keywords Bioanalysis · Ligand binding assay · Reagent lifecycle management · Critical reagent · Competitive assay · Noncompetitive assay · Assay development · Biotherapeutic · Drug development · Assay validation · Reagent characterization

8.1 Introduction

Ligand binding assay (LBA) is a generic term used to denote an assay design in which analyte quantitation depends on the intrinsic high affinity and specific interaction(s) between a macromolecular entity (e.g., usually an antibody), and an analyte of interest, often a biotherapeutic, anti-drug antibody, or biomarker. Today, ligand binding assays (LBAs) are the most widely used analytical methodology for conducting regulatory-compliant bioanalysis of macromolecules. Following their initial introduction, some 60 years ago [1], LBAs have grown steadily in use and importance within the pharmaceutical industry commensurate with the emergence and clinical successes of biotechnology-derived drugs (i.e., biotherapeutics).

As with any bioanalytical method, the goal for an LBA is to reliably provide accurate and precise measurement results for test samples. To achieve this, and especially for regulatory-compliant bioanalysis, the obligatory requisite for success is a well-developed, optimized, and validated LBA. This provides a solid analytical foundation to establish and demonstrate an assay is suitable for its intended purpose [2]. As shown below (Fig. 8.1), development and optimization are the critical first two phases for establishment of a valid LBA to support regulated bioanalysis. Even though LBAs are conceptually simple, they are deceptively complex (Fig. 8.2) and can present a barrier to success for the untrained professional. Critical for successful LBA development and optimization are a comprehensive in-depth understanding of the biotherapeutic (target analyte) and a commitment to performing iterative assay refinement and lifecycle management.

8.1.1 *Competitive and Noncompetitive Ligand Binding Assays*

Ligand binding assay designs can be categorized broadly into one of two types; competitive assays and noncompetitive assays (Fig. 8.3). These analytical designs are fundamentally different and possess unique attributes. Key differences in their characteristics are summarized in Table 8.1. A few of the design differences are noteworthy and merit additional comments. First, alternate terminology used

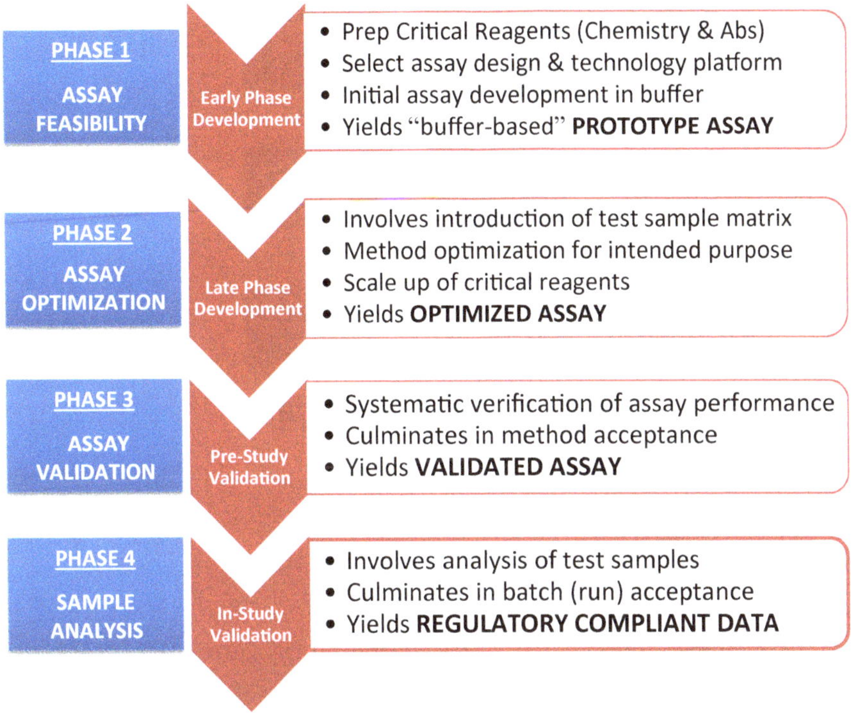

Fig. 8.1 Process flow for establishment of a valid LBA to support regulatory-compliant bioanalysis (Abs=Antibodies)

historically for LBA design has been 'reagent limiting' (competitive) and 'reagent excess' (noncompetitive). This terminology is self-explanatory and reinforces the amount of antibody needed for supporting regulatory-compliant bioanalysis. Today, because of the need to label the binding reagent (i.e., antibody), and the ability of excess reagent to augment antibody affinity sensitivity, monoclonal antibodies have evolved into being the reagents-of-choice for use in noncompetitive enzyme-linked immunosorbent assays (ELISAs). The second notable attribute for these LBA designs is the slope of the concentration–response relationship for the standard curve. All competitive LBAs (Fig. 8.3a) possess a negative slope (Fig. 8.3b), whereas noncompetitive assays (Fig. 8.3c) are characterized by a positive slope (Fig. 8.3d). Moreover, competitive assays typically display an entire sigmoidal concentration–response relationship and are fitted well by a weighted four-parameter logistic (4 PL) model algorithm, whereas noncompetitive assays often do not display a well-defined upper asymptote and can lack symmetry. Consequently, these features can make use of a weighted five parameter logistic (5 PL) model beneficial to fit standard curves for noncompetitive ELISAs. While detailed discussion of the nuances for fitting nonlinear calibration curves is beyond the scope of this chapter, some excellent references are available for supporting

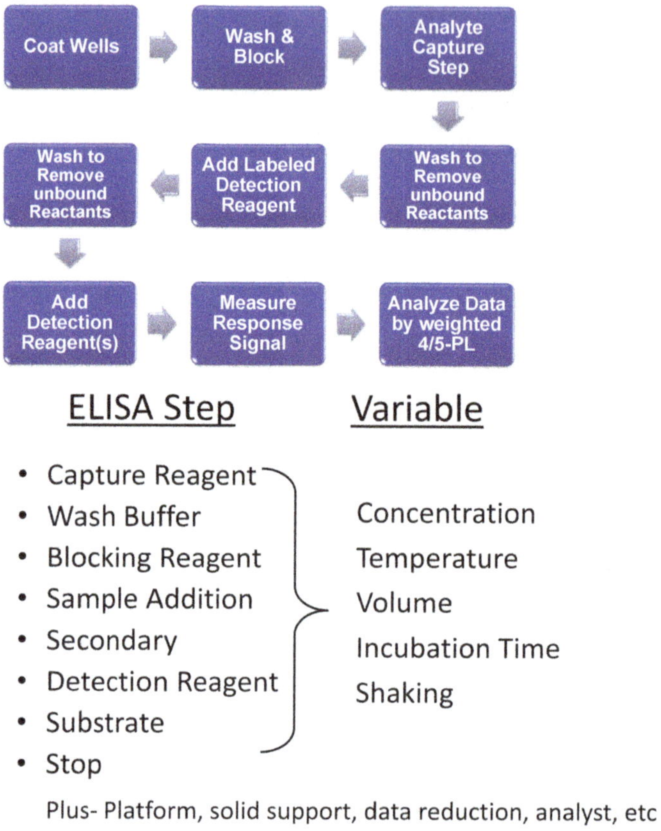

Fig. 8.2 Flow diagram for procedural execution of a typical noncompetitive LBA for regulated bioanalysis and the associated variables that require optimization

regulatory-compliant bioanalysis [3, 4]. Finally, due to the heteroscedasticity inherent in the concentration–response relationship for LBAs, use of appropriate weighting is important to ensure optimal goodness-of-fit across the entire standard curve.

The earliest assay design was the competitive LBA for which the radioimmunoassay (RIA) is the archetypical methodology. Initial description of RIA methodology was reported six decades ago by Berson and co-workers for quantitative determination of insulin concentrations in human serum [1]. For this design, measurement is performed by monitoring the extent of competition between an unlabeled analyte of interest present in a test sample and a finite amount of a labeled version of the analyte (often referred to as the *Tracer*) for a fixed and limiting concentration of a specific binder (i.e., often a polyclonal antibody). Historically, the labeled analyte was usually a radiolabeled (e.g., [^{125}I] or [^{3}H]) version of the drug. Following the liquid phase incubation, the antibody-bound and free forms of the analyte were separated by any of a variety of methods, often PEG-assisted

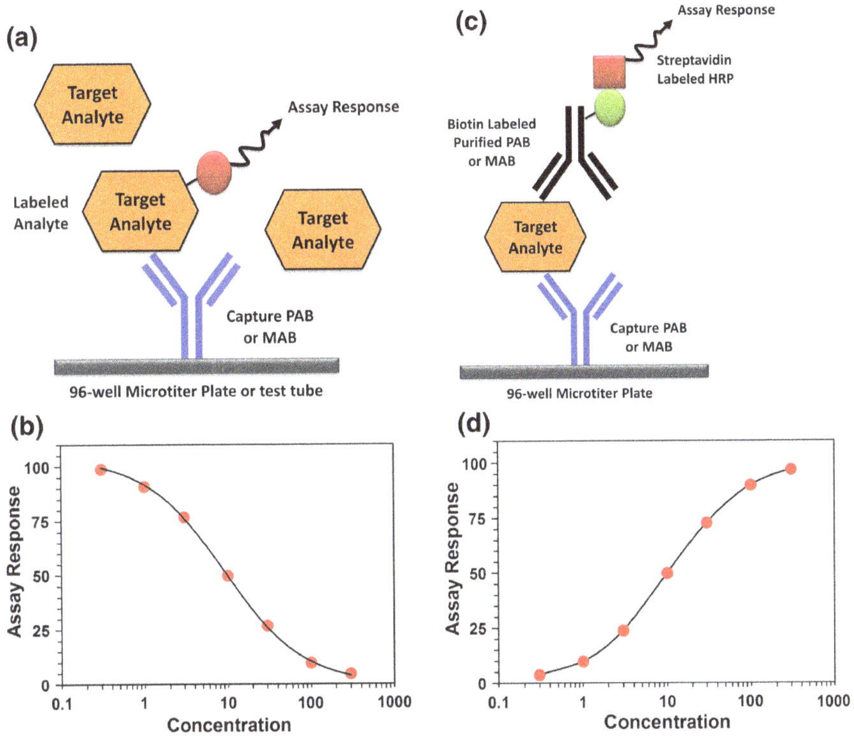

Fig. 8.3 Representative configurations for competitive and noncompetitive LBAs. **a** A standard competitive ELISA format and **b** A representative assay response-concentration curve. **c** A standard noncompetitive ELISA format and **d** A representative assay response-concentration curve

precipitation with a specific secondary (anti-IgG) antibody. The analyte concentration was then estimated by measuring the amount of radioactivity present in the antibody-bound fraction. Today, with the decreased utilization of radioisotopes in laboratory environments, radiolabeled tracers have been replaced by a variety of non-isotopic labeling strategies, such as enzymes, fluorescence, and luminescence labels. Nonetheless, the principles governing competitive LBAs today are not different from those of immunoassays used decades ago. One attractive feature of competitive LBAs is the fact that the reagent antibody is used at a fixed and limiting dilution. For example, it is not uncommon to use an antibody dilution of 1:100,000 or greater! This attribute combined with the fact that the analyte is the labeled assay component, makes competitive LBAs appealing for use with polyclonal antibodies, especially when quantities maybe limiting. It is not uncommon for a good high titer polyclonal antibody to be suitable for supporting bioanalysis of a million or more samples. Thus, with appropriate planning and execution, a validated competitive LBA method using a single lot of high-affinity polyclonal antibody can be adequate to support both the preclinical and clinical programs for a biotherapeutic. Despite

Table 8.1 Key differences in the characteristics of ligand binding assays

Assay characteristic	Competitive design	Noncompetitive design
Labeled reagent	Analyte	Antibody
Example methodology	RIA	ELISA
Antibody used	Not purified (often polyclonal antibody)	Purified (often monoclonal antibody)
Antibody dilution	High (equilibrium LBA)	Low (non-equilibrium LBA)
Binding reagent	Limiting	Excess
Standard curve slope	Negative (−)	Positive (+)
Separation of bound/free	Liquid phase (e.g., precipitation)	Solid phase (e.g., capture and wash)
Sensitivity determination	Antibody Affinity, (K_a)	Total antibody, $K_a[Ab_T]$
Analyte(s)	Both small molecules and macromolecules	Macromolecules
Incubation times	Long (e.g. 18–24 h)	Short (1–2 h)

these benefits, concerns are often expressed regarding the potential for a lack of specificity when polyclonal antibodies are used as the basis for analyte quantitation in regulatory-compliant LBAs [5, 6]. For this reason, it can be beneficial, on a case-by-case basis, to employ some orthogonal technology during method optimization (i.e., late phase development) to verify specificity. In addition to conserving reagent antibody, a few other advantages of competitive LBAs worth mentioning are the ease of development and troubleshooting of assays combined with their applicability to macromolecules (biotherapeutics) and even small molecule analytes.

Today the noncompetitive LBA design is used the vast majority of time to support regulated bioanalysis. This transition was initially driven by the introduction of monoclonal antibody technology in the late 1970s [7]. The trend has been sustained by advances in monoclonal antibody engineering and molecular biology combined with introduction of a variety of new novel non-isotopic reagents and platforms. Noncompetitive LBAs are also appealing intellectually, because the use of two distinct binding reagents directed against different antigenic sites may confer increased assay specificity for the analyte of interest. As mentioned previously, the noncompetitive assay design is appealing because it typically affords maximal sensitivity for analyte quantitation. Despite their predominance for supporting regulated bioanalysis of biotherapeutics, noncompetitive assays do present some recurrent challenges. First, the reagent antibodies must be purified and labeled to permit detection. Lot-to-lot consistency and lack of documentation for lifecycle management are known issues for commercially obtained antibody reagents [8–10]; Bradbury and Pluckthün [11]. Second, because these are reagent excess methods, noncompetitive assays can consume substantial quantities of antibodies, especially if the intent is to provide longitudinal support for both nonclinical and clinical programs. Third, noncompetitive assays require more time and resources for method development and optimization. Moreover, troubleshooting of noncompetitive assays tends to be more challenging than for competitive assays, as analytical issues can be masked as false negatives. In addition, a substantial array of assay interferences has been reported for noncompetitive assays [12–14]. Consequently, despite their conceptual simplicity, successful establishment of noncompetitive assays to support regulated bioanalysis is a nontrivial task that can consume significant time and resources, if appropriate efforts are not expended upfront for planning and early phase assay development.

In particular, the importance of high quality critical reagents cannot be overemphasized for successful method development and consistent performance of regulatory-compliant LBAs. Regardless of the assay format, LBA method performance is highly dependent on the quality of the reagents. A method that has not been optimized, but utilizes the appropriate LBA reagents, may outperform a method that has been extensively interrogated but employs the use of suboptimal reagents. As will be discussed in subsequent sections, identifying, sourcing, and maintaining high quality, well-characterized LBA reagents is absolutely critical to generating reliable bioanalytical data.

8.2 Ligand Binding Assay Reagents

8.2.1 Noncritical Reagents

When striving to develop and maintain an immunoassay that is reliable and performs consistently over time, all components of the assay system should be regarded as being potentially critical at the onset of method development. That being said, there are clearly two distinct categories of immunoassay reagents/consumables that are used by the bioanalytical scientist. Critical reagents are those reagents where either (1) there is a limited supply that is not easily replenished or (2) the reagent possesses attributes (e.g., affinity for a specific antigen epitope, a unique glycosylation profile, sample tube polymer composition, etc.) contributing to assay performance for which there is no substitution. Critical reagents are often antibodies or protein calibrators. As an example, proteins purchased as a reference standard used for the preparation of assay calibrators may be available from several sources. A quick Internet search for purified interleukin-6 protein indicates that there are more than 1300 products available. The purchase options may be segregated by species, native versus recombinant, cell expression type, physical state, stabilization, posttranslational modifications, chemical tags, etc. Clearly, substitution of one reagent for another is likely to alter the assay performance.

The second category is noncritical reagents that are those reagents that are readily available and not dependent on the use of a specific manufacturing lot or batch. It is not necessary to sequester sufficient quantities of these reagents to complete a program, as the reagents may be replenished throughout the course of the study without adversely impacting assay performance. Furthermore, the noncritical reagent often may be substituted or even generated within the laboratory. For example, the source of phosphate buffered saline may not impact the assay performance whether obtained from vendor 1, vendor 2 or properly prepared from individual components in the laboratory. That is not to say that reagent specifications should not be defined. For example, distilled water may not be able to be substituted for 18 mega Ohm reverse osmosis deionized water. As indicated earlier, it is not always apparent at the onset of method development which reagents and assay components will necessarily be critical and noncritical.

Occasionally, noncritical reagents may straddle the line between being classified as critical or noncritical. For example, bovine serum albumin (BSA) is a common buffer component that is readily available from numerous commercial sources; however, investigators can experience an unacceptable change in method performance when changing BSA sources, or even different manufacturer batch preparations. BSA is an example of a commercial reagent that may be obtained in various physical states, concentrations and purities; therefore, it is not surprising that it may be classified as a critical reagent in certain analytical methods. While it may be possible to define a reagent/consumable as noncritical, it is best practice to empirically determine the reagent status early in method development.

8.2.2 *Critical Reagents*

As discussed, critical reagents are those reagents that cannot be substituted without impacting the performance of the LBA. Obvious candidates for being critical reagents include the basic binding components of the analytical method—the ligand (e.g., reference standard of the therapeutic drug) and the capture/detection antibodies.

The generation of critical reagents has shifted in the past few decades. Historically, critical reagents were generated in-house, either to preserve intellectual property or because commercial services did not operate in a regulatory environment capable of supporting a regulated drug development program. Jump forward to present day, and it is not uncommon to outsource the procurement and management of these reagents. Drivers for this evolution in procurement of critical reagents are corporate downsizing and relinquishing of core competencies combined with a growing awareness of the importance of critical reagents [10, 15] by customers and increased emphasis on reagent characterization and quality by regulatory bodies [16].

In 2012, there were over 300 biological reagent companies. Lacking an industry standard on reagent characterization, the bioanalytical scientist must depend on the description and performance claims of individual companies for their advertised products. Some reagents are characterized only by listing the antigen used for immunization, while others may have a clone ID linking them back to specific cell lines. Most reagents will be received with a pseudo-certificate of analysis with some claim of purity and concentration. However, there is application-specific information that is necessary for determining specificity and sensitivity that is typically lacking. Usually, this information must be determined by the purchaser and requires a technical-savvy background and experience in ligand binding assay development and validation. It has been estimated that, in the United States alone, poorly characterized commercial reagents result in $350 M of wasted research spending. Moreover, there are well-documented cases where commercial reagents were unacceptable and failed to meet the most basic research needs, leading to substantial increases in costs and timelines [11]. Examples of these practices include insufficient inventory to support long-term clinical studies, unknown cross-reactivity and specificity, lot-to-lot variability, changing of clone identification labels, and incomplete or missing audit trails. With the FDA beginning to scrutinize the origins and characterization of biological reagents, utilizing a reagent of unknown origins could have long-term risk. These assay trends have led to greater awareness of the importance of early-phase method development for long-term consistency in assay performance.

8.2.3 Acquisition of Antibodies and Calibrators

Finding a source for reliable and consistent critical reagents takes due diligence. If the source is an off-the-shelf product, it is vital to understand the supply chain for that reagent. As mentioned above, if the product being purchased comes from a third-party vendor, it may be impossible to identify the true source of that reagent. For large clinical studies with multiple testing sites, it will be important to procure enough material upfront to last the entire trial. Alternatively, there is the option of procuring custom critical reagents. This avenue not only ensures the fit-for-purpose design of the reagent but also allows for continuous supply. This option can be used for generating monoclonal, polyclonal, and recombinant antibodies. The following are important considerations when searching for a vendor who supplies customized critical reagents:

- Does your supplier understand biotherapeutic drug development?
- Do they have experience in the assay design process?
- How do they characterize the products they deliver?
- Do they offer critical reagent lifecycle management?
- What is the process for reagent characterization? How do they ensure specificity and sensitivity?
- Is there a chain of custody that can be tracked back in the event of an audit?

8.3 Ligand Binding Assay Reagent Modifications

8.3.1 *Protein Chemistry and Procedures*

A unique aspect of the LBA laboratory is the requirement to manipulate biomolecules to accommodate a vast array of LBA format configurations. Each bioanalytical program is unique, often requiring the generation (rather than the commercial purchase) of critical reagents to support the project. Furthermore, there are instances where the biotherapeutic may actually be incorporated into the assay as a reagent (e.g., as a calibrator in pharmacokinetic assays or a capture reagent in anti-drug antibody assays). Compounding the challenges of supporting such programs is that the reagent may be available from a single source and may undergo slight modifications (e.g., GLP grade versus clinical grade) over the course of a formal drug development program.

Many protein/antibody manipulations may be performed on a small scale in the bioanalytical laboratory utilizing commercial kits. However, for large-scale programs or to generate consistent reagents to be used for extended lengths of time, it may be beneficial to outsource reagent management to a company (or internal specialized laboratory) that has the capability to manage these reagents over the life-cycle of a drug development program.

In general, modification of a protein/antibody requires a process that involves (1) acquisition of the required materials in a solution that is compatible with the required chemistry, (2) executing the chemical reaction, (3) cleaning up the reaction mixture to isolate the desired modified reagent away from other reaction components, (4) characterization of the modified reagent and 5) aliquoting and storage of the final reagent.

8.3.2 *Preparation for Chemical Reactions*

Biological compounds are often received lyophilized in salts or in solutions that, while stabilizing the biologic, contain additives that interfere with chemical modifications. Consequently, it is necessary to remove the interfering substance. Desalting/buffer exchange may be accomplished in one of several ways, including dialysis, filtration, chromatography, and precipitation. Dialysis uses semi-impermeable membranes to separate small molecules from larger molecules. Molecules larger than the molecular weight cutoff (determined by the membranes' pore sizes) do not pass through the membrane. Smaller molecules, ions and buffer components reach equilibrium within the dialysis membrane and the external dialysate buffer. Interfering substances may effectively be removed by refreshing the dialysate buffer.

Filtration is a very common and relatively rapid method for molecular separation based upon size. As with dialysis, molecules smaller than the molecular weight cutoff (determined by the filters' pore sizes) pass through the membrane while larger molecules are retained. Unlike dialysis, which is a relatively passive method, filtration is facilitated by gravity, vacuum, or centrifugation.

Size exclusion chromatography is accomplished by passing a reagent mixture through a bed of porous resin material. The larger molecules cannot permeate the pores of the resin and exit the bed first. Smaller molecules intercalate within resin cavities, thus delaying exit through the resin material. Collecting fractions allows for separation of different sized particles. Affinity chromatography can be used to remove either wanted or unwanted molecules from a reagent while at the same time performing buffer exchange. Finally, protein precipitation is accomplished by adding precipitating reagents (e.g., TCA, acetone, urea, etc.) to a protein mixture. The initial buffer can then be removed and the protein resolubilized in an appropriate buffer.

8.3.3 Executing the Chemical Reaction

8.3.3.1 Biotinylation

Biotinylation is the addition of biotin to another molecule preparing it primarily for bonding with avidins. Biotinylation is one of the most common protein/antibody modifications that is performed in the LBA bioanalytical laboratory. Biotinylation allows the development of a plug-and-play assay system, permitting rapid assay incubations (minutes) employing reactions that essentially go to completion. The avidin–biotin interaction is one of the strongest non-covalent molecular associations known (Kd = 10^{-15} M), is very rapid, and once established, maintains integrity under extreme conditions (pH, temperature, salt). There are numerous chemistries available functionalizing biotin thus allowing for biotinylation of a wide variety of molecules for use in different applications. Selection of the appropriate biotinylation molecule can be crucial to the outcome and should consider variables such as molecule solubility, functional target, and spacer arm length. Despite the tremendous importance of biotinylated reagents in LBAs, these reagents are ironically and inexplicably often not well-characterized by the end-user.

8.3.3.2 Reagents Used for LBA Detection

Upon capture of the ligand of interest, a mechanism for detection is necessary to quantitate the analyte concentration. It is the detection molecule that allows the user to determine the presence or absence of the antibody-analyte complex and for quantitative assays, relative levels. Detection is achieved by conjugation of a molecule to one of the assay sandwich components that either directly or indirectly generates measurable signal readout. Depending on the characteristics of the ligand, different chemistries are utilized to maximize labeling ratios and/or maintain ligand functionality. Direct conjugations include luminescence, chemiluminescence, fluorescence, or incorporation of radioisotopes on binding molecules. Indirect methods employ the use of enzymes that are capable of generating an amplified signal (often colorimetric) from a suitable substrate. As described above, biotinylation may be used to add flexibility to detection strategies. Reaction mixture cleanup is accomplished utilizing the same procedures discussed in the previous section.

8.3.3.3 Characterization and Storage

Characterization of critical reagents is a critical part of the assay development process. The assay is incomplete until material usage, storage and expiration is understood and defined for the end user or reagent generator. Regulatory agencies require that the reagent label contains a unique lot identifier as well as information on the reagent identity, concentration, storage conditions, and expiration. Physical characterizations may include pH, pKa, conductivity, concentration, purity, and ratio of detection molecule to functional molecule. Ideally, some statement on the functional activity in an established bioanalytical method is also supplied to the laboratory.

Following the development of characterization methods and specifications, stability at the desired storage conditions should be established. This can be accomplished through accelerated stability studies but should be followed with concurrent real-time examinations.

8.4 Reagent Lifecycle Management

Reagent Lifecycle Management is the process of ensuring fit-for-purpose generation and consistent supply and quality across the lifetime use of the reagent. The life cycle of the reagent begins at the moment the reagent is needed. This is where the requirements are defined. The generation of a suitable reagent requires specific input in areas such as target selection, antigen design (sometimes requires peptide syntheses or conjugation), and species selection. The process of lifecycle management should remove risks associated with changes in lot-to-lot variation, long-term stability, and supply.

Over the past few years there has been increased scrutiny placed on the determination of critical reagents used in bioanalytical methods. Regulatory agencies have taken an increased interest in ensuring these reagents are appropriately characterized and stored under defined conditions [16]. These reagents are tools being used to build novel and important medications; it is only a matter of time before there is strict guidance over how they are used, certified, and handled.

Table 8.2 The six phases of critical reagent lifecycle management

Phase	Description and examples
Reagent generation	Antigen design, target selection, screening
Reagent preparation	Scale-up, purification, conjugation, labeling
Characterization	Purity, concentration, and activity assessment
Serialization/track and Trace	Pipette into aliquots, store in vials, create unique barcodes, easy to read labels on vials, audit trail
Storage and inventory	Temperature controlled storage, inventory reports, on-going long-term stability evaluation
Distribution	Tracking of shipping, ensure shipping conditions and provide chain of custody

The process of lifecycle management can be broken down into essential phases that, when combined with long-term drug development planning, can mitigate most of the risks associated with poor quality and indeterminate supply (Table 8.2).

Of key importance in these phases is the ability to provide meaningful characterization. Within that is the capability of determining activity of the reagent. This skill requires knowledge of the end use of the reagent. For example, the reagent could be a capture antibody for a PK assay or a reference standard in a biomarker assay. To determine true activity of the reagent, there needs to be an assay developed that can represent the use of the reagent in the testing lab. To show consistency, this assay needs to be run over the course of the life of the reagent.

8.5 Assay Formats

8.5.1 Instrument Platforms

The term ‘platform’ has several definitions in the LBA world. Platform can mean the actual instrument that is being utilized to obtain a signal. It may refer to the solid surface (or lack thereof) utilized in “building” the assay. It can even be as specific as the label that generates the concentration-dependent signal that is directly proportional to the amount of protein in the sample. One other common use of the term platform, and how it will be utilized in this chapter, is in reference to the technology employed in an analytical method that produces a measureable output or signal [17]. Using this definition of a platform, one can broadly classify platforms into a few distinct categories; colorimetry, luminescence, fluorescence, radioactivity, label free, and finally those few technologies that may not fall cleanly into any single category.

The most commonly used assay platforms for ligand binding assays are colorimetry and luminescence. These two platforms utilize an enzymatic reaction that produces either color or light as an end product, depending on the substrate utilized. The most common enzymes utilized for both of these platforms are horseradish peroxidase (HRP) and alkaline phosphatase (AP). Both of these enzymes are commonly found in commercially available reagents and use widely available substrates to oxidize or hydrolyze the enzyme to the desired end product.

A more modern take on luminescence is electrochemiluminescence or electrogenerated chemiluminescence (ECL). While the end result of ECL is not different from the previously described luminescence platform, the method to generate the light is not due to an enzymatic reaction, but rather from electrochemically generated intermediates undergoing a highly exergonic reaction to produce an electronically excited state that emits light upon relaxation to a lower level state [18]. The most common label used for ECL methodologies is ruthenium, which releases photons at ~620 nm and can be detected with either a photomultiplier tube or a CCD camera [19].

Fluorescence is a specific type of luminescence, and is very similar to ECL in that the light generated is due to excitation of electrons from an outside energy source. When the electron(s) fall back into their original orbit, the resulting energy is released in the form of light. This light is the measureable signal to which a known concentration curve can be generated and used for estimation of unknown sample concentrations. The two types of fluorescence that are utilized in LBA are basic fluorescence and time-resolved fluorescence (TRF). Both of these utilize a laser to excite the conjugate to a higher state; however, time-resolved fluorescence utilizes the excitation of a lanthanide whereby the specific fluorescence signal is a long-lasting signal that can be resolved from the short lived signal of the background. This results in the potential to gain a higher level of sensitivity than may be seen using other fluorescence dyes [20, 21].

As mentioned earlier in this chapter, radioisotopes are still used both in industry and academia. Commonly used isotopes include radiolabeled iodine, [^{125}I], and tritium, [^{3}H]. While in many laboratories, the use of RIA has become passé, these radioisotopes still have their place, especially when a capture/detection reagent pair is not able to be identified or the biological analyte of interest is on the smaller side of the large molecule spectrum and not able to accommodate two binding partners reliably.

An area that has continued to generate interest is the use of label-free technologies, most commonly surface plasmon resonance [22]. This platform has long been utilized to evaluate binding affinities of reagents, although it has had a renaissance of sorts in order to evaluate a protein directly rather than indirectly.

Each of the platforms outlined have distinct advantages and disadvantages. While it is not possible to provide a comprehensive list of the pros and cons of each platform, and the decision of which platform to use for a given protein must be made on a case-by-case basis, we will try to highlight the major differentiators of these platforms here.

Colorimetric and luminescent assays are still the most commonly used platform for the reasons of cost and reagent availability. Not only is the equipment needed to run these platforms minimal, but also the consumable costs are low and the sheer number of commercially available conjugated reagents allows for an enormous number of assays that can be developed for a wide variety of protein targets. These platforms also have a few challenges. The relatively short dynamic range, typically less than two logs within the quantitative range, introduces practical challenges for measuring physiological levels in study samples [23]. Additionally, while the availability of reagents is an advantage, it can also be a hindrance because the quality of these reagents varies greatly and requires one to ensure the reagents being used are of a high quality from a reliable manufacturer. Finally, one must consider the fact that the signal is generated from enzymatic reactions dependent on temperature and the quality/stability of the substrate. While this may not be a disadvantage per se, it is certainly a consideration that must be taken into account.

ECL typically offers a much larger quantitative dynamic range that can be as high as four to five orders of magnitude. The number of manufacturers producing ECL-compatible reagents is small, and therefore the supply of the reagents can be more closely monitored. This small number of supplies is of course also a disadvantage especially if the reagent of interest is only supplied by a single vendor [24]. Another disadvantage to ECL is that it requires specialized equipment and consumables that are significantly more expensive than their colorimetric and luminescent counterparts.

Fluorescence and time-resolved fluorescence (TRF) at first appear to be the best of both worlds in that the quantitative range of the assay can be between two and three orders of magnitude and the consumables used are on the lower side of the price spectrum. Looking closer at this platform does present some disadvantages however. First, as with ECL, reading the output of a fluorescent assay requires specific equipment that has the ability to both emit an energy source and to then measure the resulting output of photons. While acquiring new equipment may not be cost prohibitive for some labs, retrofitting existing equipment may also be an option. Additionally, and specific to TRF, there is a limited number of reagents available. The most common of these is a streptavidin conjugated lanthanide that limits the assay format that can be used (e.g., detection reagents must be biotinylated and plates cannot use the biotin–avidin reaction).

There are several advantages of radioimmunoassays worth considering. First, because the assay is done in a liquid phase rather than a solid phase, the binding of reagents happens with proteins in a more native form and environment. Second, an RIA can be very sensitive and have a wide dynamic range because the readout is more digital, determined from the counting of radioactive particles hitting a detector. Third, the assay can be tuned for higher sensitivity or dynamic range based on the time a sample is in the detection chamber. There are of course disadvantages to radioimmunoassays as well. First and foremost is the fact that the materials are radioactive and require specific training and special certification from the U.S. Nuclear Regulatory Commission before any reagents can be purchased for use in these assays. Second, as with some of the other platforms discussed, specialized equipment is required to run these methods. Third, while it is possible to radiolabel proteins/peptides in a lab, this requires larger quantities of the radioisotope and someone with the skills to perform the labeling reproducibly. Finally, the reagents used for these assays are, by definition, unstable, and will need to be replenished on a regular basis or purchased just prior to performing the method. This can lead to a higher level of variability over the lifetime of the assay if precautions are not taken.

The last platform considered here is label-free quantitation of proteins. The primary advantage to this platform is the fact that the proteins of interest are not artificially modified and can be measured through direct binding to a target. Another advantage of this platform is the level of information that can be obtained, including association and disassociation constants, competition with concomitant medications, and the real-time evaluation of these traits. Disadvantages to label-free technologies are primarily the insufficient sensitivity and the low volume of throughput.

There is no "one size fits all" platform for an LBA laboratory—rather one must understand the nature of the protein being evaluated. These considerations include: the quantity that is expected to be in the test system at any given time, sensitivity requirements, cost to run the method, throughput required, and the availability of both custom and commercially available reagents.

8.6 Assay Designs

8.6.1 Assay Design

As mentioned at the beginning of this chapter, there are two basic assay designs competitive and noncompetitive assays. Noncompetitive assays can also be described as sandwich assays. By "sandwich" we mean that there is a capture reagent used to select specifically for the analyte of interest and a detection reagent that may or may not be specific for the analyte of interest. There are several variations on this theme including bridging (same reagent used for capture and detection), target capture, and monoclonal capture. Additionally, these assays can be performed in a homogeneous (capture, detection, and sample co-incubated then added to a solid surface) or heterogeneous (reagents added sequentially to the solid surface with incubations and washes between each step) format.

Below we discuss the most common assay formats used for pharmacokinetic (PK) assays, anti-drug antibody assays, and biomarker assays. Within these sections the pros and cons of the different formats will be discussed as well.

8.6.2 Pharmacokinetic Assays

Nonclinical PK studies are one of the first requirements for a regulated bioanalytical assay in a drug development program. The need to be able to confirm dosing of an animal and understand the basic adsorption and excretion profile of the drug for toxicology assessments is essential to approval of a new drug.

The first format that many laboratories utilize has been termed a generic assay, however this is somewhat of a misnomer. This assay, used for antibody therapeutics or Fc fusion proteins, uses anti-human IgG antibodies as the capture and the detection reagents. Ideally these capture and detection reagents are different from each other so as not to introduce complicating interferences or competition. Another important characteristic to consider is the species of animal the PK samples will be obtained from and any potential cross-reactivity possibilities. This assay format can be performed either homogeneously or heterogeneously, the choice of which should take the end use of the data into consideration.

There are several positives to these assays, first and foremost is the availability of the reagents from commercial sources for use on an ever-increasing number of analytical platforms. This availability allows for the bioanalyst to develop a method quickly and without the added expense of needing to have a custom, specific reagent generated. Another advantage is that this assay format can be utilized in most species to be tested in the nonclinical stage of a drug development program.

While these positives are significant, there are some very real pitfalls. As was discussed earlier, the first is the "double edged sword" involved with using commercial reagents—while the reagents are readily available, control over the quality and resupply of these reagents is lost. A second drawback to this format is the fact that this is not a method that can be used in the clinical stages of drug development because it takes advantage of human-based reagents to measure the drug in non-clinical species. Finally, these methods can be challenging to maintain as several critical reagent lot changes may occur throughout the life of the method. These critical reagent lot changes require some demonstration that the assay will continue to perform as it did with previous lots [10, 15]. A second assay format that is commonly adopted early in the life of a drug development program is a target capture method. As the name implies, this method uses the drug target as a capture reagent. This use of the drug target can provide a high level of specificity in the method. These assays can use a generic detection reagent since the initial coating reagent, after washing away unbound proteins, provides the level of specificity to reduce background and increase sensitivity of the method.

Depending on the drug target, these methods can also utilize commercially available reagents making development quick and relatively inexpensive, like the generic method described above. Additionally, because there is a level of specificity associated with the method, validation of these methods are more likely to succeed and can be used longer term. Depending on the nature of the target, these methods can be transferred for use in other species, so once the method has been developed, there is a strong likelihood that clinical method development activities may be reduced.

While it may sound like there are not many negatives to this assay format, there are several factors that should be taken into account. First one must consider the commercial availability of reagents. Most proteins available on the market today are recombinant, and have tags or Fc portions fused to them. These commercial reagents may not interact with the therapeutic well and can make finding the right detection reagent difficult depending on the origin of the Fc Fusion. Understanding the drug target is essential when evaluating this kind of format, especially if the target is a conformational epitope. Proteins binding to a solid surface such as a plate or bead are not inherent and can cause distortion of the protein. This can eliminate or greatly reduce the ability of the therapeutic to bind the target once placed onto the solid surface. The effect can be mitigated through co-incubation of the sample with the capture reagent prior to introducing to the plate or bead, but this is not always successful, because the therapeutic may not stay bound once the complex is coated onto the plate. Finally, this format requires an understanding of the drug target biology. If the target is a shed receptor or exists in large quantities in the blood, interference can be a significant hurdle to development of the method. In these instances, an additional sample clean up step or an acid disassociation step may be necessary to ensure accurate quantitation of the drug.

The third major format used in PK assays is to use custom reagents, commonly polyclonal and/or monoclonal antibodies against the drug itself. These reagents are highly specific. Depending on the strategy used to generate the reagents, a pair of reagents can be generated that not only bind the drug with high affinity, but also do not compete with each other. These are by far the most desirable as they can be used for nonclinical and clinical development, have a very high degree of specificity, and generally do not have a lot of problems with matrix effects.

While this set of reagents leads to the best, long-term method, generation of these reagents is expensive. For this reason, it is generally not advisable to look at this assay format until the program is at a stage where the assay is expected to be used for a significant amount of time. Additionally, these reagents may take longer to generate than may be acceptable for early stage, or even IND enabling studies. Finally, the expertise to generate these highly specific reagents typically does not reside with the same people developing the therapeutic or the analytical methodologies and thus will require additional resources for outsourcing, developing strategy, and program-monitoring.

As you can see, these three basic formats offer a lot of flexibility in both money and time commitment, affording discovery groups many options for use in obtaining early information about a drug candidate. As a program moves forward, a bioanalytical strategy should be developed to define critical times to make decisions on developing increasingly robust methods and the generation of custom reagents.

8.6.3 Immunogenicity Assays

Today, assays for the detection and characterization of anti-drug antibodies are one of three assay designs. These include the direct binding ELISA, the bridging ELISA, and the radioimmunoprecipitation format. Each of these designs differ in their generalized attributes and make all of them useful as part of the analytical armamentarium for supporting immunogenicity testing.

The direct binding assay design is versatile, easy to perform, does not require special equipment and represents the predominantly used format for many years. This design involves detection of antibodies after binding the target antigen to a solid surface by either passive adsorption or another strategy. The common strategy for detection of bound serum antibodies is to employ a secondary antibody that binds specifically to the Fc domain of the antibody of interest. The detection secondary antibody is conjugated chemically (such as biotin) or coupled to an enzyme (HRP or alkaline phosphatase) to enable detection by absorbance, fluorescence, or luminescence.

Despite its versatility and simplicity, the direct binding format has some limitations and is, therefore, used less often today. One major reason for the decline in popularity is the need for multiple detection antibodies. For example, one is needed to detect the surrogate antibody (such as, an anti-rabbit IgG) that is used for preparing assay controls and another to specifically detect reactive antibodies in the host's serum (e.g., anti-human IgG). Adding to the decline in the use of the direct binding assay design for immunogenicity testing has been the emergence of monoclonal antibodies as important therapeutic modalities for numerous diseases, especially chronic disorders. This category of therapeutic is not readily amenable to testing by the direct binding design, because the target analyte possesses a Fc domain just like the circulating ADA. As such, this circumstance precludes use of the secondary antibody as a common detection strategy.

Consistent with the emergence of therapeutic monoclonal antibodies has been the growth in the use of the bridging assay design. The cornerstone of this analytical strategy relies on the bivalency of the target analyte as the key structural attribute to enable detection of the drug reactive ADA. Today, the bridging design is often performed with two chemically modified versions of the therapeutic monoclonal antibody; one is typically biotinylated and serves as the capture (i.e., employs streptavidin coated microtiter plate) and the second is modified in one of a number of ways to allow specific detection. One popular strategy is to employ a ruthenium-labeled reagent to permit detection by means of electrochemiluminescence. Because detection in this format requires binding to both forms of the monoclonal therapeutic, a frequently used strategy to enhance ADA detection is to employ acid dissociation to disrupt antibody-bound complexes.

The third category of immunogenicity assay design in use today is the radioimmunoprecipitation assay (RIPA) and its variants. This approach has been used for many decades, and represents an outgrowth of the classical RIA assay to detect peptides and proteins. This design usually, but not always, employs a radiolabeled form of the therapeutic (e.g., such as [^{125}I]) that renders the technology highly sensitive and makes it suitable for use with a wide range of biotherapeutics, especially peptides. Following incubation with a serum sample, the antibody-bound and free forms of the labeled biotherapeutic are separated specifically and the radioactivity in the bound fraction is measured in a gamma counter. Despite its positive attributes, the use of the RIPA assay design has declined substantially because of the move away from the use of radioisotopes and with the dramatic growth in the use of the bridging assay design for monoclonal antibodies. Nonetheless, the RIPA remains an important option in the analytical toolbox for detection of ADA. In conclusion, ADA detection in the modern bioanalytical laboratory requires the existence of multiple design options within the analytical armamentarium so that the design and format are dictated by the physiochemical properties of the therapeutic.

8.6.4 Biomarker Assays

FDA defines a biomarker as a defined characteristic that is measured as an indicator of normal biological processes, pathogenic processes, or responses to an exposure or intervention, including therapeutic interventions. Thus, examples of biomarkers range from blood pressure to cholesterol concentration to gene sequences.

Biomarkers are measured during drug development programs to evaluate efficacy. Although many diagnostic tests are capable of measuring these analytes, the regulations that govern diagnostic testing differ significantly from the regulations that oversee pharmaceutical development. Diagnostic testing is reserved for either evaluating patient enrollment qualification or safety testing. Biomarker efficacy testing may be performed using a wide variety of methodologies, including techniques that are utilized in PK analysis such as liquid chromatography/mass spectrometry. Because many biomarkers are macromolecules composed of amino acids, LBA methods used in PK and ADA studies may be applied to efficacy biomarkers and many of the discussions in the proceeding sections apply to biomarker bioanalysis; however, a few challenges are worth discussing here.

When supporting drug development programs there are two classes of quantitative biomarker assays, Definitive Quantitative or Relative Quantitative [25]. A Definitive Quantitative assay uses calibrators fit to a regression model to calculate the absolute quantitative values for unknown samples. A Relative Quantitative assay depends upon a response–concentration calibration function. However, reference standards may not be available in a purified form that are fully characterized, or fully representative of an endogenous biomarker. The lack of a true reference standard should not discourage the team from investigating a particular biomarker, as efficacy studies are usually interested in confirming a change in biomarkers in response to a therapeutic intervention. Therefore, precision is often more critical than accuracy.

Commercial research-use-only kits are available for nearly any analyte that might be of interest to measure. However, we recommend that the pharmaceutical scientist thoroughly investigate the quality of these reagents and interrogate the kit in an internal method qualification. Furthermore, if the intended use of the data will result in FDA review, adherence to recent regulatory recommendations [26, 27, 28] is suggested.

8.7 Conclusions

LBA methods remain the most widely used bioanalytical tools for regulatory-compliant bioanalysis to support biotherapeutic drug development programs. The development and validation of a reliable ligand binding assay that is 'suitable for its intended purpose' is dependent on the availability of high-quality LBA reagents. Historically, these reagents were managed through internal resources. However, as the biotechnology community has evolved and shared best practices and as global regulatory agencies have placed increasing demands on the performance of these methods, the availability of commercial reagent services has expanded. Currently, there is a dearth of standardization with respect to developing and managing reagents to support LBA laboratories. However, knowledge and practices are rapidly increasing and internal/external partners are realizing the benefit of reagent lifecycle management.

References

1. Yalow RS, Berson SA. Immunoassay of endogenous plasma insulin in man. J Clin Invest. 1960;39:1157–75.
2. Shah VP. The history of bioanalytical method validation and regulation: Evolution of a guidance document on bioanalytical methods validation. AAPS J. 2007;9(1):E43–7.
3. Findlay JWA, Smith WC, Lee JW, Nordblom GD, Das I, DeSilva BS, et al. Validation of immunoassay for bioanalysis: a pharmaceutical industry perspective. J Pharm Biomed Anal. 2000;21(6):1249–73.
4. Findlay JWA, Dillard RF. Appropriate calibration curve fitting in ligand binding assays. AAPS J. 2007;9(2):E260–7.
5. Findlay JWA. Specificity and accuracy data for ligand-binding assays for macromolecules should be interpreted with caution. AAPS J. 2008;10(3):433–4.
6. Findlay JWA. Specificity of bioanalytical methods for pharmacokinetic support of biological macromolecular therapeutics is important. Bioanalysis. 2011;3(6):567–9.
7. Kohler G, Milstein C. Continuous cultures of fused cells secreting antibody of predefined specificity. Nature. 1975;256(5517):495–7.
8. Baker M. Blame it on the antibodies. Nature. 2015;521:274–6.
9. Bowsher RR, Nowatzke W, Sailstad JM, Khan MU. Application of commercial research-grade biomarker assay in drug development: is it time to create 'pharmaceutical-grade' kits? Bioanalysis. 2012;4(20):2427–30.
10. Rup B, O'Hara D. Critical ligand binding reagent preparation/selection: when specificity depends on reagents. AAPS J. 2007;9(2):E148–55.
11. Bradbury A, Pluckthün A. Standardize antibodies used in research. Nature. 2015;518:27–9.
12. Kricka LJ. Human anti-animal antibody interferences in immunological assays. Clin Chem. 1999;45(7):942–56.
13. Schwickart M, Vainshtein I, Lee R, Schneider A, Liang M. Interference in immunoassays to support therapeutic antibody development in preclinical and clinical studies. Bioanalysis. 2014;6(14):1939–51.
14. Tate J, Ward G. Interferences in immunoassays. Clin Biochem Rev. 2004;25:105–20.

15. O'Hara DM, Theobald V, Egan AC, Usansky J, Krishna M, TerWee J, et al. Ligand binding assays in the 21st century laboratory: recommendations for characterization and supply of critical reagents. AAPS J. 2012;14(2):316–28.
16. FDA, US Department of Health and Human Services. Draft guidance for industry: bioanalytical method validation (revised). [Online] September 2013. http://www.fda.gov/downloads/Drugs/GuidanceComplianceRegulatoryInformation/Guidances/UCM368107.pdf.
17. Spriggs F, Zhong Z, Safavi A, Darshana J, Dontha N, Kant A. et al. Ligand binding assays in the 21st century laboratory: Platforms. AAPS J. 2012;14:113–8.
18. Forster R, Bertoncello P, Keyes T. Electrogenerated chemiluminescence. Ann Rev Anal Chem. 2009;2:359–85.
19. Fang L, Lü Z, Wei H, Wang E. Quantitative electrochemiluminescence detection of proteins: Avidin-based sensor and tris(2,2′-bipyridine) ruthenium(II) label. Biosens Bioelectron. 2008;23:1645–51.
20. Hemmilä I, Laitala V. Progress in lanthanides as luminescent probes. J Fluoresc. 2005;15:529–42.
21. Yuan J, Wang G. Lanthanide complex-based fluorescence label for time-resolved fluorescence bioassay. J Fluoresc. 2005;15:559–68.
22. Daghestani H, Day B. Theory and applications of surface plasmon resonance, resonant mirror, resonant waveguide grating, and dual polarization interferometry. Biosens Bioelectron. 2010;10:9630–46.
23. Viswanathan C, Bansal S, Booth B, DeStefano A, Rose M, Sailstad J, et al. Quantitative bioanalytical methods validation and implementation: best practices for chromatographic and ligand binding assays. Pharm Res. 2007;24:1962–73.
24. Yohrling J. Ligand-binding assays: risk of using a platform supported by a single vendor. Bioanalysis. 2009;1:629–36.
25. Lee JW, Devanarayan V, Barrett YC, Weiner R, Allinson J, Fountain S, et al. Fit-for-purpose method development and validation for successful biomarker measurement. Pharm Res. 2006;23(2):312–28.
26. Arnold ME, Booth B, King L. Ray C workshop report: Crystal City VI—bioanalytical method validation for biomarkers. AAPS J. 2016; doi:10.1208/s12248-016-9946-6.
27. Booth B, Arnold ME, DeSilva B, Amaravadi L, Dudal S, Fluhler E, et al. Workshop report: Crystal City V—quantitative bioanalytical method validation and implementation: the 2013 revised FDA guidance. AAPS J. 2015;17(2):277–88.
28. Lowes S, Ackermann BL. AAPS and US FDA Crystal City VI workshop on bioanalytical method validation for biomarkers. Bioanalysis. 2016;8(3):163–7.

Chapter 9
Ligand Binding Assays in the Regulated Bioanalytical Laboratory

Johanna Mora, Charles Hottenstein and Binodh DeSilva

Abstract This chapter covers the use of ligand binding assays (LBAs) in a regulated bioanalysis environment. The various platforms employed in the analysis of protein therapeutics are varied and selection of the appropriate platform for a particular application is highlighted. The suitability, procurement and implementation of critical reagents are of major importance for LBAs and considerations for their careful selection and use are provided. In addition, various strategies, including Design of Experiments (DoE) are discussed as a means of optimizing assay conditions to produce robust and rugged assays during method development. The key parameters for the validation of LBAs are discussed in depth as well as many considerations, including automation, that can be employed when analyzing study samples in a production environment.

Keywords Bioanalytical · Ligand binding assays · Automation · Bioanalytical validation · Design of experiments (DoE) · Ligand binding reagents · LBA · Immunoassay platforms

J. Mora (✉) · B. DeSilva
Analytical and Bioanalytical Operations, Bristol-Myers Squibb, Route 206 & Province Line Rd, Princeton, NJ 08543, USA
e-mail: johanna.mora@bms.com

B. DeSilva
e-mail: binodh.desilva@bms.com

C. Hottenstein
Bioanalysis, Immunogenicity and Biomarkers, GlaxoSmithKline, 709 Swedeland Road, King of Prussia, PA 19406, USA
e-mail: charles.s.hottenstein@gsk.com

M.L. Rocci Jr. and S. Lowes (eds.), *Regulated Bioanalysis: Fundamentals and Practice*, AAPS Advances in the Pharmaceutical Sciences Series 26,
DOI 10.1007/978-3-319-54802-9_9

9.1 Understanding the Requirements of the Bioanalytical Assay

The primary objective of a bioanalytical assay is to evaluate the safety and efficacy of a drug entity administered either clinically to a human subject or nonclinically to animal species specific for the purpose of the study. Regulators have noted that the sensitivity and selectivity of bioanalytical assays are critical for the generation of accurate and precise data to support these claims. As such, guidance documents from the global regulatory agencies consistently highlight the importance of characterizing these parameters to ensure that bioanalytical method suitability is sufficient for use in a regulated study environment. In this chapter, we will describe the details of evaluating the critical parameters of ligand binding assays used to support preclinical and clinical studies in a regulated environment. We focus primarily on practical utility, as there are many publications that review theoretical and contextual aspects of this topic [1–3].

LBAs are commonly used in the quantification of protein therapeutics to evaluate a drug's toxicokinetics (TK), pharmacokinetics (PK), pharmacodynamics and immunogenicity during the drug development process. In 2015, new drug approvals reached a 66-year high, with protein therapeutics accounting for 39% of all approvals—up from 35% in 2014 and 22% in 2013 [4]. Protein therapeutics ranged from simple peptides (up to 5000 Daltons molecular weight) to more complex molecules such as antibody-drug conjugates ($\sim$150 kD molecular weight). Depending on the complexity of the protein therapeutic they may be manufactured through chemical synthesis or produced by a biological process, including the use of *E. coli* or mammalian cell lines. The current diversity of these molecules and the breath of the bioanalytical tools used in their quantification is broad. Small peptides can be quantified easily using liquid chromatographic—mass spectrometric (LC-MS) methods, while monoclonal antibody therapeutics or PEGylated molecules are quantified primarily using LBAs. Due to their complexity, antibody-drug conjugates, often require both LC-MS and LBA methods to quantify the different biological and functional components of the protein therapeutic molecule.

Given the diversity of assay needs, the bioanalytical sciences field has evolved at tremendous speed over the past two decades. The associated technology platforms have evolved alongside this growth, from straightforward radio or enzyme immunoassays to more instrument intensive platforms like LC-MS. The primary goal, during any stage of the drug development process (discovery, early development, late development or post-marketing support), is to understand the biology of the molecule, its mechanism of action, and its pharmacological impact. Once the study design and endpoints are determined, the most appropriate analytical platform —appropriate for the specific study design and able to deliver accurate and precise data—can be identified. Lastly, the analyst must consider the degree of regulatory scrutiny applied to these studies, as it will affect the level of assay validation required.

Bioanalytical support is typically required from discovery through the post-marketing phases of drug development. In the discovery phase, assays are often used to select the most appropriate candidates by understanding preliminary PK, identifying dose selection, and characterizing critical quality attributes, such as proteolytic liabilities, deamidation, disulfide exchange, aggregation, and oxidation. To produce this early-phase data, it is common practice to develop a number of unique assays for use during that development phase. Validation of these assays is not critical, as once a drug candidate is selected, a thorough understanding of the pharmacology and safety will be required. This workup will also include the immunogenicity profile for the drug, which may be important in understanding the clinical PK of the drug. At this point in the drug development process, the assays may need to be more reliable. The critical reagents (see critical reagents section in this chapter and Chap. 8) must be selected to measure the drug in the necessary form (i.e., free, total, or complexed with target) and demonstrate the necessary assay robustness that aligns with the level of regulatory scrutiny required for the phase of drug development. As the data become more important to supporting a drug approval filing, assay validation must become more thorough, so all pertinent regulatory requirements are met. We highly recommended that the reagents used to optimize the assay during this period are scaled-up and appropriately characterized for long-term use in the regulated environment.

Starting from the first GLP designated toxicity study, assays must be validated for their intended use, and the reagents secured, must be in sufficient quantities to support the duration of the initial studies. This is the point in drug development , where the first attempts to develop a robust assay that can be transitioned to support all anticipated future clinical activities is performed. It is preferable to have the same assay format used during the preclinical toxicity studies, transitioned and validated to support the clinical studies. Many times, however, assays require redevelopment to meet the sensitivity requirements of the clinical studies, to reduce potentially high run failure rates or to qualify a new lot of reagents.

Once the therapeutic is considered safe and efficacious for its intended treatment of disease, the LBAs need to be robust enough to support large-scale clinical trials over a long period of time. At this stage, there are many facets of the LBA's life cycle that need to be considered, including: (1) procurement of the reagents in bulk, (2) transfer of the assays to third party vendors, (3) a strategy to monitor the progress and performance of the assays and (4) expansion of the assays' use into additional diseases that were not initially considered at candidate selection (Fig. 9.1).

The bioanalyst's role over the course of the drug development process is one area overlooked in many publications. A bioanalytical scientist who initially supports the drug's discovery may develop a deep scientific understanding of the assay and its capabilities, knowledge that must translate to future applications of the method. As the drug development process enters the regulated phases, focus often

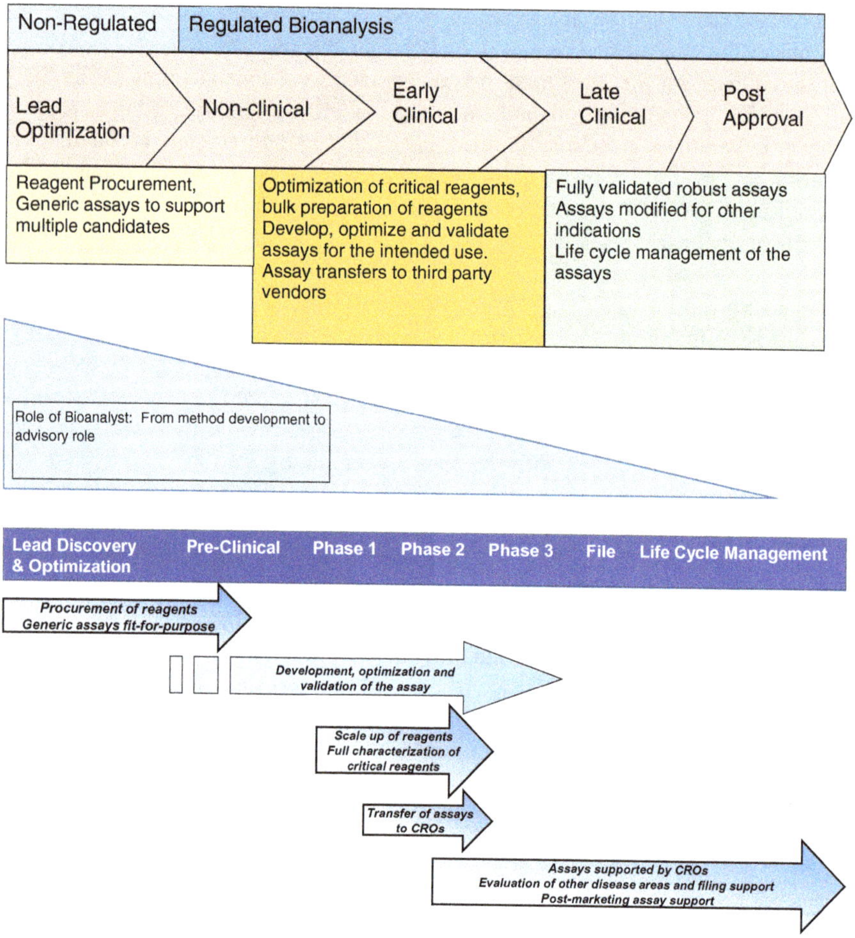

Fig. 9.1 Schematic of the stages of drug development that ligand binding assays support and the role of the bioanalytical scientist

shifts to more procedural, regulatory, and assay robustness issues, while preserving technical considerations. At this point, the validation of a well-developed assay becomes more critical. Additionally, as the drug proceeds through to the later clinical stages, the bioanalyst may assume yet different roles as an advisor, data reviewer, method troubleshooter, and project manager. Alternatively, and depending on the business model, these latter roles may be assumed by a different group of individuals either in the same department or even by a completely different organization, such as a Contract Research Organization (CRO). When a different organization assumes these duties, it is critical to keep the connection between the original method developer or validation scientist and the newly assigned staff. This connection promotes a seamless communication and information flow, enabling flawless execution of the method.

9.2 What Do We Measure in a Bioanalytical PK Assay?

Therapeutic proteins have unique characteristics very unlike their small molecule counterparts. Most are highly complex and can act more specifically, allowing them to function in a more focused manner. Commonly available therapeutic proteins (e.g., MAbs, replacement proteins, peptides) are generally well-tolerated with very few side effects. These facts are reflected in the more than 250 approved protein therapeutics currently on the market [4].

It is very important for the analyst to keep in mind the mechanism of action of the protein therapeutic, (i.e., binding to cells, membranes or circulating target proteins, and their complex structure) in order to determine which form(s) of the drug may exist in the biological system once administered. As depicted in Fig. 9.2 therapeutic proteins can be free (not bound to the biological target) and pharmacologically active, or complexed to its biological target (e.g., receptors, soluble targets, antidrug antibodies). In addition, there may be active or inactive fragments of the therapeutic resulting from enzyme metabolism. Bioanalytical method development needs to incorporate measurement of the most relevant form of the drug in order to understand the overall pharmacologic activity and/or safety of the molecule (i.e., free, fully complexed with target, partially complexed with target, or total). For example, during preclinical drug development, a knowledge of the drug's molecular forms present in circulation and at the site of action aids in understanding the safety of the molecule. However, the most relevant molecular form related to drug safety is often measured by developing a total-drug assay. In later stages of development, a PK assay that is capable of measuring the circulating unbound (free) therapeutic is typically preferred. For more detail, we refer readers to the numerous publications on this topic [5–7].

For the purpose of this chapter, the most important practical aspect of understanding a protein therapeutics' molecular form is to identify a suitable assay format and the necessary reagents to be generated. Typically, it is very difficult to measure precisely the free therapeutic in the presence of soluble circulating target, and this challenge is described in more detail later in this chapter (*Reagent Pairing to Measure Desired Therapeutic Form*). It is the bioanalyst's responsibility to communicate to the project stakeholders the benefits and limitations of the proposed assay format.

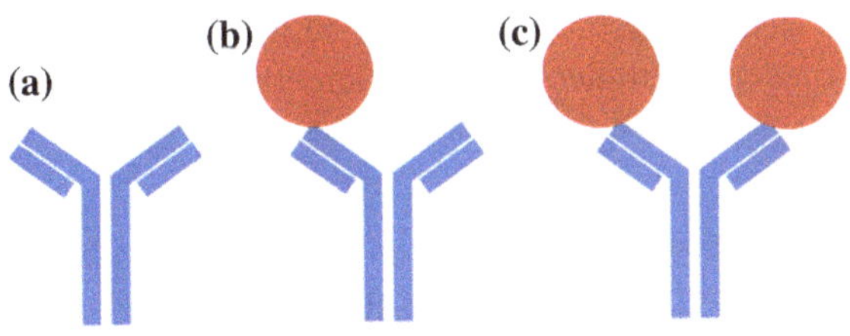

Fig. 9.2 Cartoon illustration depicting the different molecular forms that may exist in the biological system for an antibody therapeutic. The antibody depicted is a bivalent molecule. **a** unbound, free or the pharmacologically active molecule. **b** Partially bound molecule. **c** Total or fully bound/pharmacologically inactive molecule

9.3 Technology Platforms

Traditional immunoassay platforms such as RIA (Radioactive ImmunoAssay), EIA (Enzyme ImmunoAssay), and ELISA (Enzyme-Linked ImmunoSorbent Assay) are the workhorses in the LBA discipline. The use of RIAs, while exquisitely sensitive, are less common in today's bioanalytical laboratory due to financial, regulatory, and logistical issues related to the safety and disposition of radioactive material and the advancement of alternative technologies and techniques. The early regulated assay format for insulin was an RIA method that provided the data for its development as a major biotherapeutic medicine [8]. ELISA and EIA platforms are commonly used today with different variations of enzymes, labels, and detection report tags. The electrochemiluminescence detection platform primarily distributed by Meso Scale Discovery (Meso Scale Discovery, Gaithersburg, MD) has grown significantly in the past decade in the LBA field. The original bead-based flow technology developed by ORIGEN Corporation was converted to a plate-based system with capture reagents coupled directly to a carbon electrode. A number of variations of this format, with a range of immobilization chemistries, are commercially available and allow for a good deal of versatility when developing bioanalytical methods in the regulated laboratory. The Gyrolab™ immunoassay system (Gyros AB, Uppsala, Sweden), is another workstation that has been adopted in the regulated LBA space and provides high throughput with suitable sensitivity while using lower volumes of samples and reagents compared to traditional microtiter plate-based platforms. The Gyrolab is a fully automated platform where microfluidic technology is coupled with fluorescence detection. The capture reagents are biotinylated for ease of immobilization onto micro columns packed with streptavidin-coated beads. The columns are part of the microfluidic structure within a compact disk, where all of the immunochemical reactions take place. The reagents and sample flow through and interact with the micro columns by centrifugal force. A major caveat to keep in mind when working with this platform is the near absence of incubation times. This necessitates higher quality reagents that can capture the analyte of interest effectively in a very short period.

Several LBA-based emerging technologies with the potential to be used in the regulated bioanalytical environment have been identified and evaluated by the AAPS Emerging Technologies Focus Group. What follows are brief descriptions of a subset of these technologies, as they cannot be fully described in the context of this chapter. We refer the reader to publications [9–13] on this topic for further details on their potential use in regulated bioanalysis, sensitivity, and potential challenges.

- The Erenna® immunoassay system (Singulex Inc., Alameda, CA) is a bead or plate-based system where an immunocomplex is formed in a plate or bead between a detector molecule with a flourescence tag and the analyte. The fluorescence-tagged detector molecule is then eluted off the immunocomplex in a very small volume and subsequently read on a single molecule counting device. This "pre-concentration" of the detector molecules in a small volume is one of the factors aiding in the instrument's ability to quantify analytes in the picogram to femtogram per mL range.

- The AlphaLISA® (amplified luminescent proximity homogeneous assay) platform from Perkin Elmer is a homogeneous non-wash proximity immunoassay that utilizes donor beads conjugated to phthalocyanine; phthalocyanine converts O_2 to singlet $^1O_2$ (oxygen with a single excited electron). If an acceptor bead is within ~200 nm proximity of $^1O_2$, the latter can transfer its energy to thioxene derivatives within the acceptor bead, resulting in light production at 615 nm. One of the main reasons it has not been wildly adapted for use for regulated assays is that costly customized beads are required to perform the method. Similar to the Gyrolab workstation, AlphaLISA methods have shorter assay incubation times and wider dynamic ranges than typical colorimetric ELISAs and require lower sample volumes.
- Similar to Singulex, Simoa is a single molecule immunoassay platform, which is microparticle-based technology developed by Quanterix (Quanterix Corporation, Lexington, MA). It is based upon the binding of the analyte to specific antibodies coated onto paramagnetic beads. The immune complexes are labeled with an enzyme capable of generating a fluorescent signal. The complexed beads are isolated in 50-fL chambers, and fluorescence imaging is used to detect single protein molecules. Its sensitivity is in the double digit femtogram per mL range with the advantage of multiplexing by encoding the beads with dye and mixing different bead types together in an assay. In addition, each encoded bead type is conjugated with a unique capture antibody to facilitate multiplexing.
- The ViBE Workstation from BioScale (BioScale is now called Proterixbio, Inc., Billerica, MA), uses acoustic assay technology based on an oscillating membrane where an immune complex containing the analyte is captured for detection. The membrane is sensitive to mass changes that affect its vibrating frequency in a manner proportional to the analyte concentration. The instrument allows for walkaway operation after the initial sample preparation. Most of its applications are for biomarkers; in our hands, successful development of both pharmacokinetic (PK) and immunogenicity (IMG) assays have been achieved, however, the IMG method had higher variability than normally observed with plate-based assays.

If during the drug development process there is a need to switch analytical platforms, a cross-validation plan that incorporates the use of study samples and QCs should be executed, in addition to the basic analytical parameters evaluated during method validation. This approach provides the highest confidence when evaluating data across studies, even if supported by different platforms [14–16].

9.4 Critical Reagents

It has been highlighted by many [17–19], as well as in Chap. 8 of this book, that the functional attributes and structural integrity of reagents used in ligand binding is critical to the performance of the assay. As defined by O'Hara et al. [17], critical reagents are typically produced via biological processes, which may lead to variability as different lots are prepared. Also, the noncovalent interactions between the reagents and the biotherapeutic change the labeling, aggregation patterns, and variability in their production, potentially impacting the assay performance over the life cycle of the bioanalytical method. O'Hara et al. [17] is a valuable reference outlining the many types of reagents, their generation, selection, maintenance and inventory. The purpose of this section, then, is to touch on some of these points from a practical perspective.

The most common reagents used for therapeutic drug capture and isolation from the measured matrix typically include the biological drug target and anti-idiotypic (ids) antibodies. While these materials are commonly used to capture and isolate analytes from matrix, either of these reagents can also be used for detection. However, the most common reagents used for detection are anti-ids and antihuman Fc antibodies specific to the IgG subclass of the biotherapeutic antibody. Other antibodies, used for detection purposes, are those that bind to constant framework regions other than the binding domains or complementary determining regions (CDRs). Anti-VKappa, anti-adnectin, and anti-PEG for PEGylated therapeutics are good examples of more generic common framework-based detection reagents. Finally, the biotherapeutic drug itself is also a critical reagent. The final format of the assay (homogeneous, sequential, or competitive) and the decision of which reagents to use may depend on three main factors: (1) availability; (2) which therapeutic form of the drug needs to be measured; and (3) desired sensitivity and quantitation range. Other items critical to the performance of the assay are addressed under the method development section of this chapter (*Selection and Optimization of Buffer and Reagent Conditions*).

9.4.1 *Reagent Availability and Procurement*

It is wise to consider reagents available commercially as it may save time and reduce costs. However, it is very important to outline a strategy for the characterization and qualification of new lots of these reagents. For example, an agreement with the vendor on the type of assay used to release the material and the acceptance criteria may help avoid problems with sequential lots of the material. There are situations where vendors have not been able to provide a second lot capable of reproducing the results observed with the original lot. The white paper from King et al. [20] and Chap. 8 provides some helpful points regarding reagent availability and procurement. In cases where no commercial reagents are available or they do not fulfill the assay requirements, it is critical to initiate a custom reagent generation campaign at the time a lead drug candidate is identified. The generation of anti-idiotypic antibodies against the drug, from immunization to antibody characterization, is a process that takes 6–9 months on average, so any delays in the start of the process will impact future method development activities. It is also important to have clear communication with the group(s) generating the custom reagents and performing the purification to ensure timelines stay on track. Depending on the workflow, it is possible that efficiencies are lost if the group or analyst that screens the produced supernatants (monoclonal, mAb) or test bleeds (polyclonal, pAbs) is not available to perform the analysis promptly to enable decisions on which clone to scale up or when to perform production bleeds. It is also desirable to implement standard practices across all drug development programs and to create groups focused on process oversight and reagent characterization. The creation of such a group provides added efficiencies, but also provides confidence in the quality of the reagents obtained commercially or produced by custom reagent generation campaigns.

9.4.2 *Common Methods for Reagent Evaluation*

One of the platforms with the highest throughput for reagent suitability screening is the Octet® (ForteBio, Menlo Park, CA) and hence one of the primary platforms of choice for supernatant and custom reagent screening. However, previous reports [21, 22] observed that specificity and binding on the Octet platform did not necessarily translate into good reagents for LBA methods being developed on the Gyrolab workstation or the MSD platform. This may be explained by the assay format. For the Gyrolab workstation, higher affinity antibodies are typically required as the contact time of sample and reagents with the streptavidin bead bed is in the order of seconds. Based on this understanding, it is recommended to narrow the selection of clones to a manageable number by first screening on the Octet and subsequently evaluating those reagents on the platform of choice for method development [23]. If the platform is changed later in the development process, the reagent selection may need to be revisited and a cross-validation between the two methods performed, especially if the change happens on or after the use of the assay in the clinical setting. Another important point to consider is that PEG molecules may interfere with the interferometry technology that ForteBio uses for biomolecular interaction measurement. Thus, reagent pairing and titration experiments for PEGylated compounds should be performed only on the platform of choice for method development.

Figure 9.3 shows a common and practical way to screen mAbs generated against the drug. Depending on the screening format, anti-species antibody against the Fc region of the custom mAb reagent is immobilized on a microtiter plate or bead-based solid-phase, which is then be incubated with the therapeutic drug to measure binding. In the case of the ForteBio platform, this is the last step. For other platforms, such as Gyrolab or MSD, the drug introduction is followed by incubation

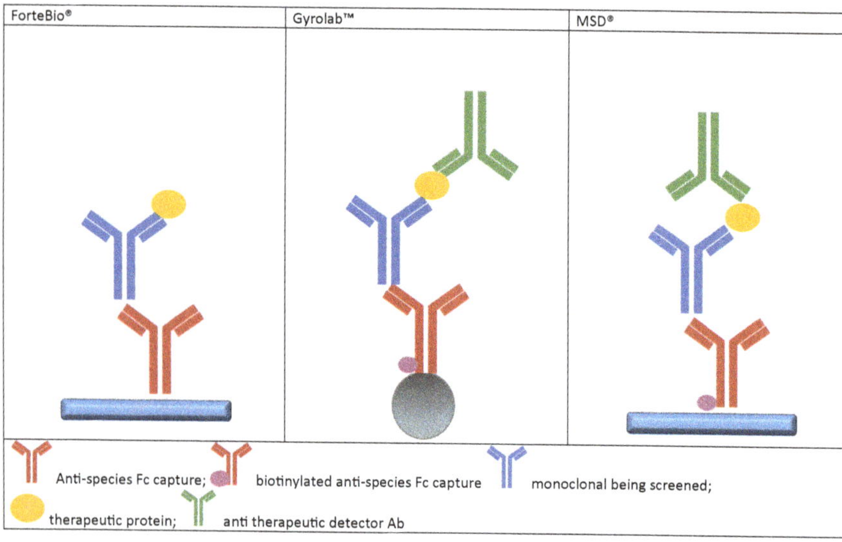

Fig. 9.3 Common platforms for screening monoclonal antibodies generated against a therapeutic

with labeled target or a labeled antibody already known to bind to the drug. Another approach to screen custom mAb reagents is to immobilize the therapeutic drug target onto the selected screening platform solid-phase. When using this approach to screen reagents for suitability, it is important to consider how direct immobilization of the therapeutic drug target to a polystyrene or other solid phase surface may affect the target's conformational integrity and its ability to bind to the mAb reagents being screened.

9.4.3 Importance of Reagent Consistency and Crossover

Case studies have shown where clone contamination resulted in the failed qualification of a new lot of reagent which in turn delayed the support of sample analysis during a custom reagent generation campaign [24]. In this situation, the initial lot that was scaled-up was not characterized. Prior chromatographic analysis of the material would have indicated the presence of two clones versus a single clone in the initial lot. In the second lot prepared, only one clone was present and a new reagent scale-up was required to avoid delays in sample analysis. Haulenbeek and Piccoli [18] reported on a study involving the redevelopment of a biomarker assay for an endogenous soluble protein to meet new sensitivity and drug tolerance requirements. During redevelopment of the assay, three previously tested mAbs, were reevaluated. All three of the original lots of monoclonal antibodies were relabeled with Meso Scale Discovery Sulfo-Tag and tested in the original assay format. Two out of the three antibodies produced lower responses than expected.

Subsequent characterization of the incorporation ratio by mass spectrometry and re-labeling confirmed that a lower than-expected incorporation ratio was the root cause of the less than optimal assay response. The authors explained that the labeling of lysine is a random process, usually resulting in a Poisson-like distribution with a varying number of labels per antibody or protein. This would mean that an incorporation ratio of one to four labels per molecule will likely have a significant amount of unlabeled protein, explaining the poor signals for the lot with low incorporation ratios. Based on this observation, it is typically recommended to test ratios of 10:1 (label:protein) and 20:1, depending on the protein size in order to obtain at least seven labels on the protein.

Further reagent characterization may consist of analysis by BiaCore (GE Healthcare Bio-Sciences, Pittsburgh, PA) or Mass-1 (Sierra Sensors, Greenville, RI) to provide information on the affinity to target and potentially aid in epitope mapping. A new platform in this space is the SQlite (SQI, Canada), which is an automated microarray immunoassay platform that enables dual characterization of antibodies [9]. For example, different epitopes may be printed onto the array and if desired, up to five isotypes of the antibody may be captured and simultaneously identified in a subsequent step. This type of approach would be typically most relevant to support immunogenicity assays. An integrated Hamilton liquid handler can automate the process with a minimum serum volume of 10 μL. The main limitation of the SQlite is that it is a closed system that requires the method development and the printing of arrays to be performed by SQI Diagnostics.

Once the reagent scale-up and labeling conditions are established, it is more efficient to prepare larger (25–50 mg) lots of material for method development, validation, and sample analysis to allow for consistency and avoid tag-to-protein ratios variations if scaled-up at different times. However, proper caution should be taken if the tag is unstable and large quantities will be stored and used over long periods of time. In situations with unstable tags, multiple lots should be tested, if possible, during the method validation process.

9.4.4 Inventory and Reagent Sustainability

Assay life cycle management refers to the bioanalytical method maintenance required over the period that the method is used for study support. This can often cover sample analysis during extended or multiple studies that could span years. For programs in mid to late stage clinical development, with a large number of studies that extend over long periods of time, generation of new reagent lots, and their qualification monitoring via assay performance is important to ensure the consistency of the data within and across studies. For organizations overseeing the development of several therapeutic drugs, there is benefit in centralizing activities related to reagent inventory and characterization, as it is faster to implement a best practice approach within a smaller centralized group. In addition, there is efficiency in communication when there is a single point of contact with collaborators. When handling reagents and their characterization across many programs, there is often workflow consistency that can benefit from the use of automation. Such activities include: (1) the aliquoting of reagents, (2) labeling and uploading into inventory, and (3) lot preparation for characterization. Establishing collaborations between the reagent group and the automation team benefits the organization.

Another important activity for programs in late phase clinical development is the transfer of hybridomas to stable Chinese Hamster Ovary (CHO) cell lines for production of reagents. This will ensure that support of bioanalysis for large study sample numbers will never be at risk due to lack of reagents. It is also common for groups outside of the bioanalytical organization to be interested in these reagents for other sample assessments such as receptor occupancy assays or sample pretreatments to remove drug interferences. Initiating this activity prior to late phase clinical drug development is not common as in many cases the assay may be redeveloped or optimized to meet new requirements [16, 25]. For example, lower administered drug dosages may require improved assay sensitivity or new regulatory expectations may justify assay redevelopment to meet the refined criteria. In such cases where new reagents and method are used, cross-validation between assays is important to ensure that the data between assays and across studies correctly characterizes the clinical pharmacology profile of the biotherapeutic protein drug.

9.5 Selection and Optimization of Buffer and Reagent Conditions

Once platform selection and LBA format have been determined based on reagent quality and availability, the bioanalytical method must be optimized and evaluated for suitable analyte selectivity and sensitivity. A well thought out experimental design is critical to the optimization of any bioanalytical method and ultimately leads to the best performance parameters and the overall level of method robustness. In general, the majority of current immunochemical bioanalytical platforms contain a solid-phase component, which is typically the starting point for reagent condition optimization. In addition, when optimizing reagent conditions for any bioanalytical method, the general practice is to optimize a single reagent or method component at a time, starting from the solid phase and working upward through the assay. Advances in automation have increased the method optimization using DoE as will be described later in this chapter. In a typical immunoassay format, the first step is to optimize the capture reagent conditions, followed by the selection of the most appropriate blocking solution. Sample conditions such as the sample diluent and MRD can then be examined followed by the detection and development steps. Between each step in the immunoassay format, removal of unbound material occurs via a washing step. Even with a finite number of reagents or components to evaluate, it may become resource prohibitive to examine every one, and it is often common practice to begin with a standard format when initiating method optimization experiments. This is especially true for assay buffers and sample diluents.

In contrast to the more common stepwise heterogeneous LBA formats, homogeneous LBA assays such as FRET (Fluorescence Resonance Energy Transfer) present their own unique set of challenges often requiring simultaneous optimization of all components and reagents involved in the method. These types of assays have no stepwise isolation and removal of excess reagents or endogenous interfering substances, making reagent optimization that much more critical compared to heterogeneous methods. Another common homogeneous LBA format first incubates the capture and detection reagents with the sample in solution to form the LBA complex prior to being introduced to the solid-phase. Once in proximity to the solid-phase, the plate is washed and read. To facilitate this type of homogeneous format, the capture reagent is often biotinylated and must be optimized to the capacity of the streptavidin solid-phase to obtain the best performance.

9.5.1 Buffer Selection

The buffers used for immunochemical methods can be broken down into a few distinct categories organized by purpose. These categories consist of coating buffers, blocking buffers, washing buffers, samples diluents, and detection reagent buffers. Often the basis of these buffers can be identical or very similar between categories with minor differences specific to the buffer function. The predominant form of sample diluents and detection reagent buffers used for immunochemical methods consist of buffer salts, such as sodium phosphate and sodium chloride, which contain a carrier protein or bulking protein and a nonionic surfactant. The buffer salts are tailored to maintain a certain level of buffering capacity and mimic physiological conditions, typically around 10–100 mM at pH 7.4 for sodium phosphate and 150 mM for sodium chloride. Carrier proteins, the most common of which is bovine serum albumin (BSA), can range between 0.1 and 1.0% by volume and act to reduce nonspecific binding to both matrix components and vessel surfaces. Nonionic surfactants are also used to reduce nonspecific binding events by disrupting weak, nonspecific interactions with unrelated matrix components. Of the numerous nonionic surfactants available, polysorbate 20 (Tween 20) is the most common. Phosphate buffered saline containing BSA and Tween 20 (PBST) is the most common and widely used buffer for reagent and sample dilution, alternative buffer salts, carrier proteins, or nonionic surfactants include, tris (hydroxymethyl), aminomethane (Tris), casein, gelatin, and triton X-100. Tris is commonly used as a replacement for sodium phosphate and brings the added feature of blocking free amines that may be present in the matrix from not fully purified reagents. Casein and gelatin can be used to replace BSA as the principal carrier protein in the buffer. And lastly, triton X-100 is a suitable alternative nonionic surfactant for Tween 20. Although these types of buffers can be used for blocking steps to minimize nonspecific binding (NSB) by increasing the carrier protein concentrations to approximately 3–5% by volume, commercial blocking solutions are readily available and often provide better inhibition of NSB events. Different to most reagents, coating, blocking, sample diluent, and wash buffers can vary widely depending on the capture mechanism, means of immobilization, and matrix being examined.

9.5.1.1 Coating Buffers

The primary difference between coating buffers, used to passively adsorb capture proteins, and reagent, blocking, or sample diluent buffers is that coating buffers lack any form of carrier protein or nonionic surfactant. These substances will prevent the hydrophobic interactions between the capture protein and the surface of the solid-phase required for immobilization. Biological buffers best suited for the purpose of protein immobilization to a polystyrene solid-phase, typically contain a higher than physiological pH closer to the capture proteins isoelectric point (pI), which aids in driving passive adsorption. For instance, the pI of an antibody can differ greatly, ranging from 6.5 to 9 depending on its amino acids sequence, but for most antibodies is approximately 9. For the effective immobilization of most proteins, a suitable buffer with the capacity of maintaining pH 9.5 is often a sufficient starting point. One commonly used coating buffer for the passive adsorption of proteins is 50 mM sodium bicarbonate at pH 9.5. It should be noted that when immobilizing the assay capture mechanism by the means of passive adsorption, it is important to maintain conformational integrity of the capture protein to guarantee sufficient analyte binding. This is typically confirmed experimentally on the particular platform being used and evaluated by examining response on a fit-for-purpose scale. If the conformational integrity of the capture protein is compromised by passive adsorption to the solid-phase, then additional strategies may be required for successful immobilization of the unique binding antigen. The most common strategy is biotinylation of the specific capture protein, which often allows for immobilization while maintaining conformational integrity. There are numerous commercially available biotinylation reagents with various conjugation chemistries and linker arm lengths, allowing for another level of optimization. These biotinylated capture reagents can be immobilized to a streptavidin or neutravidin treated solid-phase thereby functionalizing the solid-phase. The coating buffer choice can vary and range from the use of a commercial blocking buffer to PBST depending on the specific solid-phase and the impact on assay performance. Lastly, the addition of buffer stability components to reduce fungal growth may be considered, depending on the frequency of buffer use, the required shelf life of the buffer and the various storage condition options available. Some common examples of such preservatives or biocides are sodium azide, which was widely used with radio-labeled immunoassay buffers for preservation. As a result of its toxicity and ability to inactivate the most commonly used non-radiolabeled immunochemical reporter tag, horseradish peroxidase; sodium azide was quickly replaced with other suitable alternatives. One of the more common alternatives are isothiazolone-based preservatives such as ProClin® 300.

9.5.1.2 Sample Diluent

Selecting the most appropriate sample diluent can enhance method sensitivity, selectivity, and improve overall performance. In most cases, standard assay buffer solutions used for reagent dilution such as PBST will be more than suitable for use as a sample diluent, though depending on the matrix and required sensitivity, alternative sample diluents may be needed. A common approach when selecting sample diluent alternatives is to evaluate the suitability of commercially available blocking buffers. These buffers often contain a variety of carrier or bulking proteins at high concentrations, thus providing additional blocking capacity in the presence of the sample matrix. If overwhelming the matrix with a mixture of high concentration protein does not improve assay selectivity or sensitivity to desired levels, then additional reagents may be needed to mitigate endogenous matrix components that could be interfering or competing for analyte binding. Although the sample diluent is not the only type of reagent providing benefits to a bioanalytical method, it is a good starting point. The addition of mouse plasma (or serum) (approximately 3–5% by volume) into the sample diluent can act as a free scavenger and reduce or eliminate specific cross-linking events between the capture protein and the species-specific matrix. This is especially true if the capture protein is mouse derived, allowing potential interferences in the matrix to bind to the mouse plasma added and not the capture reagent. The addition of mouse plasma or species-specific matrix at similar concentrations to those listed previously, can also be introduced to the blocking buffer, to aid in reducing specific and nonspecific binding events. In addition to the numerous unknown endogenous matrix components that may cause issues for an immunoassay method, many known interferences can be resolved by fully understanding the biology of the target, the disease state, and the pharmacology of the protein therapeutic. One such interference is heterophilic antibodies which are weak, multi-specific antibodies produced against poorly defined antigens often arising from social or occupational exposures [26]. While heterophilic blocking reagents are commercially available, they may not be necessary, since species-specific plasma added to one or more of the bioanalytical assay buffers, is often sufficient to block heterophilic antibody binding, thereby reducing or eliminating interference and often improving the selectivity of the LBA.

9.5.2 Minimum Required Matrix Dilution

LBA are often developed to measure the analyte in matrix without extraction. This is especially true for therapeutic proteins. One of the more critical reagent parameters requiring optimization is the percentage of sample matrix to be used in the immunoassay method. This is typically assessed by dilution of different amounts of sample into diluent buffer. Although samples can be analyzed without dilution to maximize sensitivity and minimize sample manipulation, assay selectivity is often compromised. As a result, sample matrix dilution is an important step to reduce or eliminate subtle matrix differences between individuals, allowing for acceptable sample selectivity. To achieve acceptable sample selectivity, the dilution of sample matrix using sample diluent is optimized so that acceptable assay selectivity is obtained among individuals. The degree of dilution required (expressed as a matrix percentage) is known as the minimum required dilution (MRD) and can vary depending on the assay platform and quality of reagents. Over the course of method development, the need for additional sample dilution will depend on the assay performance and observed interferences among individual lots of matrix or subjects. Although it is ideal for the MRD in matrix to mirror the performance of a matrix free sample such as buffer, matrix MRDs that do not match a spiked buffer sample can still be robust and acceptable as long as the required selectivity and recovery is obtained. In more extreme situations where there is no suitable matrix available, a surrogate matrix may be used for calibration and possibly quality control (QC) purposes. Most surrogate matrixes are typically buffers that closely resemble the authentic matrix but are free from endogenous analytes. A key qualifier for using a surrogate matrix is the ability to demonstrate parallelism between the surrogate and authentic matrices [27, 28].

9.5.3 Reagent Optimization

Once the initial assay format is established, including starting reagent buffers and sample diluent selections, optimization can be performed for the remaining steps of the assay. This is typically performed by titrating a single reagent at a time while all other steps in the method remain constant. This is referred to as a single titration and as previously mentioned, it is often the best approach to optimize from the solid-phase step forward. The capture reagent conditions should be evaluated first, by titrating various concentrations of the capture protein to determine the optimal binding of the analyte that produces the required sensitivity and range of the bioanalytical method. To understand and optimize the ideal conditions for the capture reagent, preliminary analyte calibration standards are included with calibration points above and below the targeted range of the assay. Titration of the capture reagent typically includes a matrix blank to assess the baseline response for each concentration of capture reagent being examined and the experimental signal to noise (S/N) of the matrix blank to each calibration standard, providing an estimate of the method's sensitivity at each coating condition.

Reagent titration should be performed for each step of the immunoassay method. Since reagent optimization is not a static process, a reevaluation of optimal conditions may be necessary for previously optimized immunoassay steps. One way to avoid the potential need to go back and reexamine already optimized conditions, is to evaluate multiple reagent conditions at one time. This practice is referred to as cross-titration and involves the evaluation of typically two reagents against each other in the same experiment to identify the optimal conditions from the reagent combinations examined (Fig. 9.4). This type of approach is often used when optimizing the

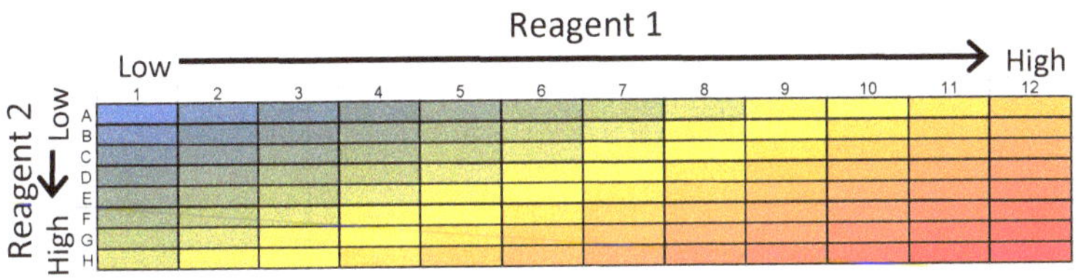

Fig. 9.4 Cross titration plate map

conditions for the detection step where there may be more than one detection step in the assay format. When optimizing primary and secondary detection steps, such as detection reagents labeled with streptavidin followed by a biotin reporter tag, it is often beneficial to perform these types of experiments by cross-titration of the primary and secondary detection reagents in combination, to identify the ideal conditions for the combined reagents. Additional opportunities to capitalize on the benefits of optimizing reagents in combination by cross-titration will depend on the individual immunoassay format, but it is a useful approach to consider whenever possible.

9.5.4 Selecting the Means of Detection

The means of detection or reporter tag is typically specific to the platform being used such as Ruthenium for the Meso Scale Discovery (MSD) or Alexa Flour® 647 for the Gyrolab™ xP workstation. There are, however, a number of immunoassay platforms capable of switching between multiple reporter tag label technologies. All of these multiple reporter tag label technology platforms are microtiter plate based. Examples of these are the EnVision® from PerkinElmer and the SpectraMax® M from Molecular Devices. These platforms are able to read a number of reporter tag technologies, such as absorbance, fluorescence, luminescence, and time-resolved fluorescence (TRF). This allows the selection of the best reporter tag system for any particular immunoassay. For instance, benefits in sensitivity or dynamic range can be achieved by selecting one reporter tag over another. In addition, buffer solutions for reporter tag development are typically paired with the reporter tag and platform being utilized. If the required assay sensitivity cannot be achieved by the evaluation of various platform and reporter tag options, then the potential benefits of reporter tag amplification can be pursued. Reporter tag amplification can be achieved by maximizing the number of reporter tags conjugated to the primary detection reagent. If maximizing the reporter tag ratio alone does not provide the required amplification then the use of a secondary detection reagent can be examined to increase LBA sensitivity. Secondary detection reagents have the potential to increase the overall S/N since theoretically they would bind to multiple tags on the primary detection reagent. For instance, if the primary means of detection is a monoclonal mouse antihuman IgG conjugated to digoxin; the benefits of secondary anti-digoxin reporter tag detection (Fig. 9.5a) can be compared to detection by

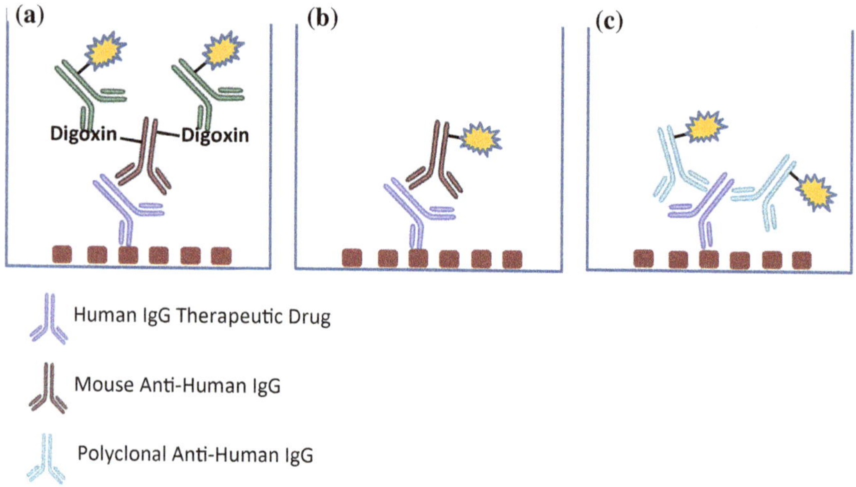

Fig. 9.5 Reporter tag amplification strategies. **a** Representation of a secondary anti-digoxin reporter tag detection; **b** Direct reporter tag conjugation of the primary monoclonal mouse antihuman IgG antibody; **c** Polyclonal primary detection reagent directly conjugated to the specific reporter tag

direct reporter tag conjugation of the primary monoclonal mouse antihuman IgG antibody (Fig. 9.5b). Alternatively, a polyclonal primary detection reagent, directly conjugated to the specific reporter tag (Fig. 9.5c) may generate a similar S/N response to the secondary anti-digoxin reporter tag format.

9.5.5 *Reagent Pairing to Measure Desired Therapeutic Form*

There are a number of considerations involved when selecting suitable ligand binding assay reagents such as the proposed assay complex or format, the purpose of the assay (i.e., what is the preferred form of the analyte being measured) and the overall level of regulatory quality required. Often, some level of actual or implied functional integrity is preferred when selecting reagents for pharmacokinetic (PK) and pharmacodynamic (PD) LBA. The most basic example of this type of LBA format would be to incorporate the molecular target of the therapeutic protein as the means of capture and immobilization with detection via a distal molecular moiety. For protein therapeutics with more than one mechanism of action, such as monoclonal antibodies with effector functions or bi-specific products, reagent selection for the bioanalytical method should attempt to incorporate all of the relevant biological activities present in the analyte.

Modifications of the drug, such as deamidation, oxidation, and aggregation generated through in vivo processes have the potential to exhibit different PK and PD functions and may affect safety and/or efficacy of the molecule. Therefore, distinction between the administered protein and the modified in vivo populations can be an important factor to consider when selecting reagents for immunochemical bioanalytical method development. Ultimately, reagent selection should be based on understanding the potential for the presence of in vivo molecular variants and allow for the measurement of the desired protein therapeutic form based on this knowledge. For instance, if the analyte of interest is a humanized monoclonal antibody specific for an antigen that exhibits a high circulating concentration, the questions would be: "*Are we interested in quantifying the free unbound fraction of the antibody, or do we want to measure the total concentration of circulating drug regardless of the amount complexed with target?*" Depending on the answer to this and other similar questions, the reagent selection process could potentially take two different paths. If the free unbound fraction is of the greatest interest, then evaluating various forms of the drug target as a capture reagent might be the best path forward. Some of these capture options may include the native or recombinant forms of the drug target as well as target fragments that contain the specific binding moiety for which the humanized monoclonal antibody is specific. Alternatively, if the total amount of circulating drug is the most desired analytical form to be measured, then a reagent that is specific for the therapeutic antibody, yet is more generic and does not target the functional binding region as the means of isolation from the matrix should be considered.

9.6 Design of Experiments (DoE) for Optimizing Assay Parameters

Traditionally, LBAs are developed and optimized using one reagent or step at a time as described in the above section. The authors feel that this chapter would be incomplete without a brief description of the DoE. DoE is an efficient and systematic statistical process used to plan experiments and assess the effects or interactions that multiple factors may have on the measured assay response. With DoE it is possible to efficiently assess multiple factors (e.g., the effect of coating, incubation times and detector concentrations) in one experiment. The advantages of testing several factors at once include an understanding of the interactions of each factor, robust assay development and fast turnaround time [29]. Assay responses that could be monitored during DoE testing include, but are not limited to, signal-to-noise ratios (S/N), standard curve parameters and instrument response at specific analyte concentrations.

A few complexities need to be mitigated when performing DoE experiments. To reduce uncontrollable variables such as the effect of well position on the plate and shaking effects (if any), the position of samples on the microtiter plate should be randomized. Also, once the data is obtained, it is necessary to calculate the statistical variance of each parameter so that the interpretation of the data is meaningful.

In our experience, it has been easier to split the process into three phases: design, execution and analysis. DoE designs may have distinct goals, including: screening (used to explore many factors and their influence on the response and their appropriate range), optimization (has the goal to predict response for all combinations of factors and the optimal experimental point) and robustness testing (has the objective to ensure that the method is tolerant of small changes). The experimental design itself can be further classified into classical and custom designs. Classical designs are well described in many textbooks [30, 31] and can be set up using basic software such as Excel. Common classical designs include fractional factorial and Plackett–Burman for screening, and central composite (see example in Table 9.1) and Box–Behnken for optimization. Alternatively, custom designs may combine elements of screening and optimization. In these cases, it is easier to utilize software such as JMP® to plan experimental runs. Within JMP, the user defines which factors they want to include in their design, the interactions between these factors that they wish to examine, and the responses they will use to assess the effect of the assay factors. After entering this setup information, the software provides the plate layout for each assay step (Fig. 9.6).

Automation platforms, such as Janus, Tecan, Hamilton, or other liquid handling systems are beneficial to employ during the execution phase as they help reduce errors introduced when tracking plate positions manually and when different variables/factors are being tested at once. Based on the liquid handling platform used, appropriate procedures are scripted for each of the experimental steps. Once the execution phase is complete, data analysis is performed using statistical analysis tools such as GraphPad Prism or JMP®. The latter provides the ability to create an empirical model for the assay that can be used to obtain predicted response values when the factor settings are modified (Fig. 9.7)

Table 9.1 Schematic of a three factor central composite design

	1 Capture	2 Secondary	3 Detector
Stock concentration (mg/mL)	5.0	2.0	1.25
Volume (mL)	50	50	50
1	100	200	50
2	200	400	100
3	300	500	150
4	400	600	200
5	500	200	100
Units	ng/mL	ng/mL	ng/mL

Fig. 9.6 Example set up of a plate map for a DoE experiment

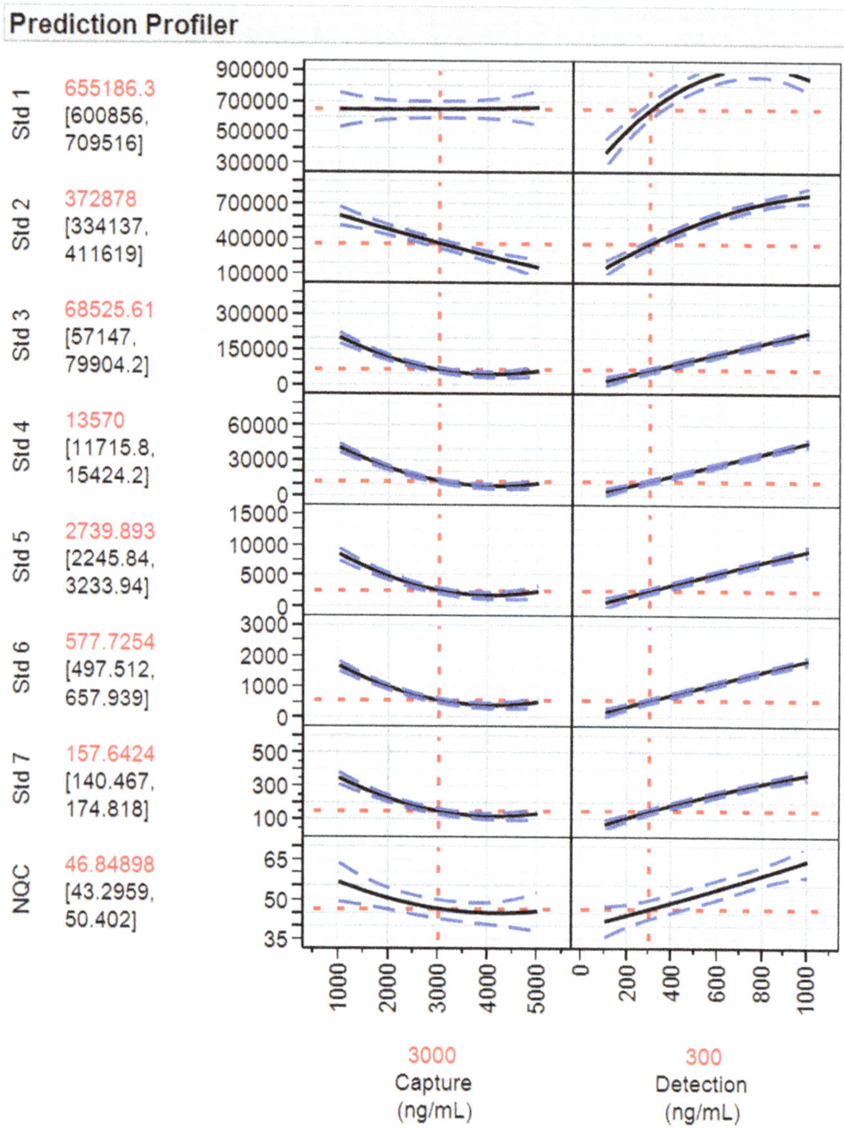

Fig. 9.7 An example of a prediction profiler from JMP® software

9.7 Evaluation of Critical Ligand Binding Assay Performance Parameters

Once buffer and reagent conditions have been selected and optimized, an initial evaluation of the critical performance parameters should be conducted, to achieve a baseline understanding of the method's robustness and if the performance is

Table 9.2 Critical ligand binding assay performance parameters

Performance parameter	Description
Calibration point and regression algorithm selection	Examine calibration data and select simplest most appropriate calibration fit
Precision and accuracy	Determine preliminary precision and accuracy estimates
Selectivity	Perform preliminary examination of matrix selectivity
Linearity of dilution	Perform preliminary examination of dilution linearity
Tolerance to common interferences	Perform preliminary examination of known or potential interferences

suitable for the assay's intended use prior to initiating formal qualification or validation exercises. A list and brief description of these exercises are presented in Table 9.2. The reader may find more detailed descriptions by DeSilva et al. [3].

9.7.1 Calibration Point and Regression Algorithm Selection

The ultimate goal when selecting calibration points and determining the optimal fitting algorithm is to obtain the best possible accuracy and precision across the maximum usable calibration range. For immunochemical LBA, the measured response compared to the analyte concentration is generally nonlinear and sigmoidal in nature, with the lower and upper asymptotes residing near the respective background and maximum measured responses of the method (Fig. 9.8). The use of nonlinear calibration is somewhat different than other commonly used analytical methods such as chromatographic assays, which typically employ a linear model for calibration fitting. In addition, the overall calibration ranges for LBA are typically narrower than chromatographic assays. The most common nonlinear fitting

Fig. 9.8 Basic components of nonlinear calibration curve

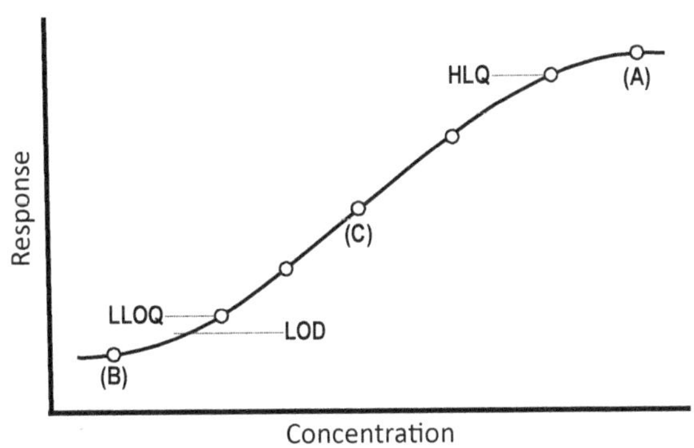

algorithm used for ligand binding assay calibration is the 4-parameter logistic (4-PL) regression. However, if the calibration curve is asymmetric, the nonlinear fit may require an additional parameter to account for the asymmetry [32]. In this case, the use of an additional asymmetry parameter may benefit the overall accuracy and precision as well as increase the maximum usable calibration range resulting in a 5-parameter logistic (5-PL) regression. It should be noted, that for all nonlinear fitting algorithms, the variability is generally greater at the upper and lower asymptotes versus the inflection point, or geometric center, for LBA that employ 4-PL or 5-PL calibration curves (Fig. 9.9). With this being the case, individual calibrators may demonstrate inferior sample accuracy and precision at the outer asymptotes compared to the center of the calibration curve, where the concentration and response may exhibit a more linear relationship (Fig. 9.9). The inclusion of weighting into the calibration curve fitting regression often results in improved performance at either the lower ($1/x$ or $1/x^2$) or upper end ($1/y$ or $1/y^2$) of the calibration curve. For LBA, calibration curves should contain from six to eight nozero calibration standards that are evenly placed in a logarithmic fashion across the full quantitative range of the bioanalytical method and analyzed in duplicate for each calibration concentration. Anchor concentrations above and below the quantifiable range of the method can be considered if it improves the precision and accuracy of the nonlinear fit at the upper or lower asymptotes. The physical placement of calibration standards throughout an immunoassay run should sufficiently account for variability across the concentration range of the assay. Traditional statistical evaluations for goodness of fit such as correlation coefficients are not good indicators to determine the acceptability of nonlinear fits. The preferred method for determining nonlinear goodness of fit involves evaluation of the accuracy of back-calculated calibrator concentrations from the nonlinear calibration line of best fit. For LBA, this value is typically ±20% of the nominal calibrator concentration and can be as great as ±25% for the calibrators at the upper and lower asymptotes of the standard calibration curve. Whereas the lower limit of

Fig. 9.9 Typical variability of a nonlinear calibration curve

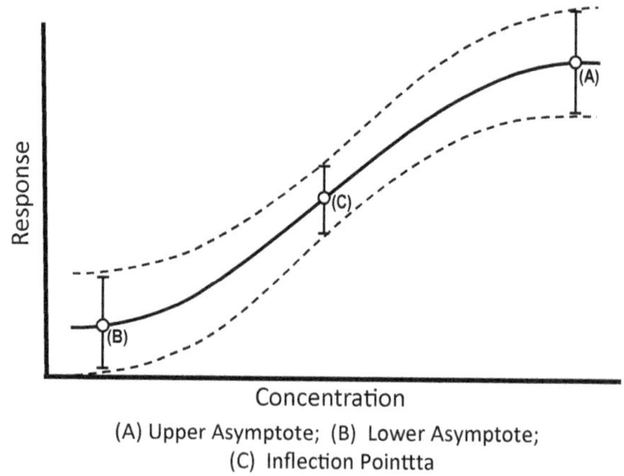

quantitation (LLOQ) is typically based on the ability to measure the analyte to a specific accuracy and precision tolerance criterion, the lower limit of detection is estimated based on the variability of the baseline response from the measurement of blank matrix. Typically, the concentration calculated from the response at 2–3 standard deviations from the average blank matrix response is how the lower limit of detection is estimated. Lastly, any method's calibration data is only as good as the calibration reference material supplied, which can vary greatly over the course of drug development and should always be considered when evaluating the quality of the data produced. A standard practice for assessing this quality is to evaluate the recovery of QCs prepared with a second lot of material against a calibration curve prepared with the original reference lot.

9.7.2 *Characterization of Precision, Accuracy, and Selectivity*

A rudimentary understanding of method suitability can be achieved early in the development process by characterizing the general level of method accuracy, precision, and selectivity. The selectivity of a ligand binding method is ultimately based on the specificity of the reagent pairs chosen for the capture and detection of the therapeutic protein being measured. Ideally, the best reagent pair is one that demonstrates no cross-reactivity with undesirable therapeutic protein variants or other endogenous sample matrix constituents. To characterize the overall selectivity of a bioanalytical method, multiple sources of matrix should be evaluated both neat and fortified with the analyte to ensure accurate recovery as well as confirm the lack of quantifiable response due to specific or NSB from blank control matrix. If during the course of these assessments, selectivity proves to be unacceptable, one of the first approaches that might be examined to gain acceptable selectivity would be to adjust the MRD by lowering the percent matrix in the assayed sample. Though this will likely improve method selectivity to acceptable levels, the sensitivity and the overall calibration range of the bioanalytical method may be negatively impacted. If adjusting the MRD by lowering the percent matrix in the assayed sample does not resolve the observed selectivity issues, then reexamining the various steps of the ligand binding assay is a proven strategy. This includes changing the blocking or bulking buffer to a solution with a higher concentration of total protein to provide additional blocking capacity; including additional wash steps; increasing thc overall salt content of the wash buffer to reduce specific low affinity cross-reactivity and normal NSB events; and adding reagents (e.g., heterophilic and IgG blockers) to mitigate endogenous matrix components that could be interfering or competing for analyte binding. If all of these efforts fail to produce suitable assay selectivity, then a reevaluation of reagent pairs, or examining the potential to extract the sample by acid dissociation or other means can be considered depending on the size and nature of the therapeutic protein. In addition to characterizing and understanding the bioanalytical method selectivity, the preliminary accuracy, and precision should also be evaluated. The accuracy and precision of the method should be assessed by establishing a number of controls with nominal concentrations that span the full range of the calibration curve, including the LLOQ and upper limit of quantitation (ULOQ). Typically, three controls in addition to the LLOQ and ULOQ controls will be sufficient to adequately cover the full quantifiable range of the method. The controls should be prepared from a separate aliquot of drug reference material than was used to prepare the calibration standards, and measured in the assay in a single run to determine the accuracy. The controls should initially be examined as single replicates and depending on the level of reproducibility (% CV) observed, the benefit of additional replicate sample analysis can be further examined if required.

The assay accuracy is the between-run percent deviation of the analytical results from the nominal concentration of each control. This is typically expressed as the % bias or % recovery. In addition to accuracy, the between-run imprecision [% RSD (% CV)] for each control assayed should be determined. The % bias and % RSD should be ideally less than 20% for each determination with a total % error (% bias plus % RSD) for each quality control concentration of less than 30%. For robust ligand binding methods the % bias and % RSD can typically be well below 10% while for more difficult assays a % bias and % RSD of less than 30% can be completely acceptable depending on the application and purpose of the data [33].

9.7.3 Evaluation of Dilution Linearity and Potential for Prozone Effects

Dilution linearity is defined as the ability to accurately dilute samples with concentrations above the analytical calibration range of a method into the measurable range of the assay. The ability to dilute samples accurately into the quantifiable range of a LBA can be hampered by a number of factors such as matrix interference or the saturation of analyte relative to the reagent concentrations and MRD of the sample matrix. LBA methods that employ homogeneous incubation, where the capture and detection reagents are added simultaneously and compete for a limited number of binding sites are more susceptible to analyte saturation than stepwise LBA methods. For either homogeneous or stepwise LBA methods, reagent saturation due to high analyte concentrations, results in inhibition of the ability for the reagent(s) to bind to the analyte and produces a nonlinear biphasic response termed the prozone or "hook" effect (Fig. 9.10). High concentration samples that exhibit prozone effects result in sample concentrations being inaccurate and under reported. To experimentally determine the concentration at which acceptable dilution linearity is obtained and beyond which a prozone effect may occur, matrix samples are prepared at the highest possible concentration ("prozone assessment samples") based on the available analyte reference material and diluted into the quantifiable range of the assay and analyzed. The reference material in the prepared prozone

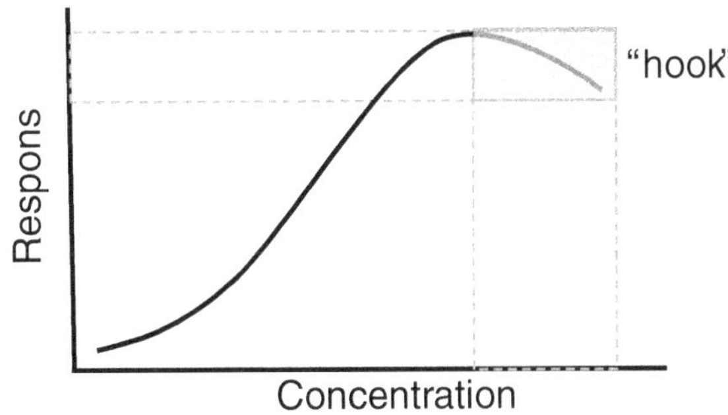

Fig. 9.10 Prozone "hook effect." The plot illustrates an analyte response curve where response saturation has occurred and demonstrates response inhibition at concentrations beyond the upper asymptote of the calibration curve

samples is typically limited to 5% by volume, which places a limit as to how high a concentration can be examined experimentally. This may or may not be comparable to expected study sample concentrations. The results of the prozone assessment samples diluted into the assay range should yield results that meet the assay acceptance criteria discussed above for precision and accuracy. If the study sample concentrations are expected to be much higher than the experimental concentrations examined during the prozone assessment, then in-study sample dilution (i.e., sample parallelism) will need to be performed on the actual study samples to demonstrate the lack of prozone effect, thereby confirming data quality. If sample parallelism experiments from the actual study samples demonstrate a prozone effect, further experiments will need to be performed to determine what level of additional dilution is required to resolve the prozone condition and achieve suitable data quality. This is typically performed by serially diluting the highest concentration study sample and observing the resulting concentrations. The concentrations of prozone-exhibiting samples will increase as the dilution factor increases across the serial dilution range employed in the experiment, until the prozone effect disappears, producing a plateauing of sample concentration results. Once the additional dilution required to resolve the prozone effect is identified, all study samples with concentrations above those determined not to have prozone issues must be subsequently reanalyzed at the qualified additional dilution. Alternatively, if the method is in a state of early development, increasing the LBA capture reagent is an option to provide additional capacity, and allow greater tolerance to analyte saturation and prozone causing inhibition. Lastly, certain LBA platforms, such as the MSD and Gyrolab™ xP workstation platforms, lend themselves to broader dynamic ranges (i.e., 2–3 logs or greater) implying a greater analyte capacity. This translates into a better tolerance to prozone effects.

Thus, understanding the dilutional integrity of an assay when dealing with high concentration samples; the consequence of prozone effects based on capture reagent capacity; and the known LBA platform limitations or benefits can aid the bioanalytical scientist in developing and optimizing the most reliable and robust methods possible.

9.7.4 Interference Tolerance Testing

While LBAs are flexible and often highly specific, they can suffer from interferences due to unknown and unpredicted components within the sample matrix as a result of their dependence on reagents. The presence of interference in LBAs can result in falsely elevated or low results and can lead to the misinterpretation of data. There are two distinct types of interference relative to LBAs, analyte-dependent or specific interference, and analyte-independent or nonspecific interference. Analyte-independent interferences refer to nonspecific matrix constituents that function independent of analyte concentration. In the case where the matrix is a blood product (plasma, serum, etc.), hemolysis and high lipid content are two good examples of potential analyte-independent, nonspecific interferences. To examine LBA tolerance to these types or other blood product interferences, hemolyzed (approximately 2%) and high lipid content (greater than 300 mg/dL) matrix are obtained or experimentally produced for use as interference test samples. The samples should be examined both fortified with the protein therapeutic being measured and unfortified to determine the presence or lack of false positive or false negative results. Any observed response of the non-fortified test samples should be no greater than the response of the method at the LLOQ, and result in a non-quantifiable measurement for the level of interference in the test sample to be tolerable. For the fortified interference test samples, the same assay precision and accuracy acceptance criteria discussed previously should be met. If the interference test samples fail to meet the defined acceptance criteria at the measured level of interference, then reselection of an appropriate MRD or an alternative suitable sample preparation technique should be considered.

Analyte-dependent interferences directly interact with one or more of the reagents in the LBA and include compounds with structural similarities to the measured protein therapeutic that cross-react with the assay reagents. One of the more common causes of this type of interference includes protein therapeutic cross-reactivity from soluble targets or molecular isoforms, which may either be inherent to the system biology or enhanced as a result of a particular disease state. Increasing immunoassay tolerance to soluble targets in the measured matrix can be handled using a number of common immunochemical techniques. These include acid dissociation, the use of acidic sample diluents for sample dilution, or by increasing the salt content in the wash step(s) of an assay to remove interferences with weaker binding affinities than those of the protein therapeutic and the immobilized capture reagent. Interference from soluble circulating target or target isoforms may not always be viewed as an interfering substance, but may need to be considered as part of the natural biological system that can complex with the drug. Understanding this relationship usually requires more than one assay, such as free and total target, to provide information on the levels of these different forms.

The potential for molecular cross-linking is another reagent-specific interference from matrix components that can occur by forming a bridge between the LBA reagents and endogenous matrix components causing a falsely elevated concentration. Some of the potential culprits that might result in molecular cross-linking between LBA reagents are endogenous proteins that circulate in some multimer form (dimer, trimer, etc.) or binding proteins that can bind to the LBA capture protein and cross-link with the detection reagent. A good example of molecular cross-linking is the case of a therapeutic antibody specific for oncostatin M (OSM). When OSM was used to capture the therapeutic antibody and subsequently detected by an anti-IgG reagent conjugated to a reporter tag, the LBA response demonstrated an elevated signal at all concentrations examined. After investigation, it was determined that the cross-linking of an endogenous circulating protein such as Glycoprotein 130, which can bind to OSM and contains IgG like domains was the likely cause of the interference. To resolve the observed interference, an idiotypic antibody specific for the binding region was generated and used to replace the OSM protein as the LBA capture reagent thereby eliminating potential interference from endogenous cross-linking proteins.

Other potential causes of LBA interference are immunogenicity to the administered therapeutic protein and drug–drug interactions from co-administered medicines. LBA interference caused by immunogenicity can be broken down into antidrug antibodies (ADAs), autoanalyte antibodies, and heterophilic antibodies. ADAs are antibodies that are produced as a result of humoral immunity from a recognized foreign body, typically a large molecular weight protein administered to a mammalian biological system, which can interact directly with the measured analyte. ADAs have also been referred to as anti-therapeutic antibodies (ATAs) or human antihuman antibodies (HAHAs). ADAs can be either analyte neutralizing or nonneutralizing and have the ability to reduce the measurable analyte concentrations in the tested matrix through in vivo clearance or by interference in the LBA method. ADAs can also sustain the levels of the therapeutic protein in the circulation. Neutralizing ADAs have the ability to specifically bind to the epitopes responsible for the biological or pharmacological activity of the analyte while nonneutralizing ADAs heterogeneously bind to nonactive portions of the analyte. Either form of ADA has the potential to result in in vivo clearance changes or LBA interference. Depending on the format of the LBA, neutralizing ADAs typically have a higher likelihood to interfere with the LBA method. Another factor which can contribute to LBA interference is the maturity and titer of the ADAs present in the sample. In general, low-level transient ADAs are often insignificant in comparison to the circulating therapeutic protein concentration, while high level, persistent ADAs may be more likely to impact in vivo therapeutic protein clearance or result in LBA interference. During the course of the method development process, it is important to understand and characterize the potential for ADA interference. This is often difficult, since ADAs are highly heterogeneous and can vary significantly from one sample to the next. Therefore, preliminary characterization of ADA tolerance is commonly determined using a surrogate ADA to determine the potential for interference. Suitable ADA surrogates can vary widely depending on the analyte, but are commonly polyclonal antibodies derived from hyperimmunization of a preclinical species such as rabbit or goat to mimic the heterogeneity that might be observed with endogenously derived antibodies. By using a surrogate ADA, LBA tolerance can be determined as discussed earlier by examining the recovery of analyte-fortified and unfortified samples with various levels of surrogate ADA. This allows for a preliminary understanding of LBA ADA tolerance that can later be correlated with the actual ADA tolerance of the method once ADA positive samples become available. Translation between the surrogate ADA tolerance and the actual ADA tolerance can then be examined and the results applied to the interpretation of future sample analysis.

Heterophilic or autoanalyte antibodies are another type of immunogenic material that have been shown to demonstrate a lower prevalence for LBA interference than ADA immunogenicity and can be present in both normal and disease populations [34, 35]. These types of antibodies are naturally formed endogenous antibodies that preexist prior to the administration of the therapeutic protein by an immune response to an alternate substance, which has common cross-reactivity to an epitope of the measured analyte. Heterophilic or autoanalyte antibodies are typically of low affinity and have poorly defined specificity which can usually be mitigated to some degree by a number of commercially available substances, such as heterophilic blocking reagent (HBR) or immunoglobulin inhibiting reagent (IIR) [36, 37]. Interference from co-administered medications with therapeutic proteins are typically less of an issue than with small molecule drugs, unless the co-administered drugs have the potential to directly interact with the analyte or the immunoassay reagents, or if the therapeutic protein is a cytokine or a cytokine immunomodulator. Therapeutic proteins that are cytokine modulators have the potential to modify the metabolism of small molecule drugs, which influence the expression or stability of specific cytochrome P450 enzymes and drug transporters. Conditions where it is most important to examine the potential for interference from co-administered drugs include combination cancer therapies. Since therapeutic proteins as a whole do not typically undergo meaningful metabolism or transport, the potential is limited for small molecule co-administered drugs to cause interference. However, therapeutic peptides such as exendin-4 have the ability to convert from an agonist to an antagonist or demonstrate increased potency upon metabolism [38]. To understand the potential risk for interference from co-administered medication on therapeutic proteins or the quantitative LBA, the biology and pharmacology of therapeutic proteins, the specific disease being treated, and which cytokines regulate which P450s in vivo must be considered [39]. When considering the potential impact from any of the aforementioned causes of LBA interference, the use of both dilutional parallelism and spiked recovery experiments are universal in their ability to identify and confirm interference from all above-mentioned sources.

9.8 Automation in Ligand Binding Assays

The topic of automation in bioanalysis for LBA has been discussed by many [40–42]. Across companies, there seems to be consensus regarding its value in the discovery and drug development stages to support studies with larger sample numbers. However, not all its applications may be justified by a return on investment (ROI). When establishing the best workflows for the implementation of bioanalytical methods, it is also important to consider that there are some significant differences between automation of regulated bioanalysis for small versus large molecule drugs (e.g., interfacing to LIMS, instruments, creation of linear vs. 8×12 array sequence files, etc.). Therefore, adequate support from automation users with expertize beyond the liquid handler is required for the success of any automation endeavor. While many companies support bioanalysis without any automation, there is a more prevalent trend to justify its implementation beyond efficiency gains and increased throughput [41]. Such justification includes the positive impact it has on the safety of the workforce (via reduction of repetitive motions that may result in injuries and reduced exposure to biologic matrices) and enhanced assurance of the quality and fidelity of the data being generated. The reader is referred to other papers [43–48] that apply automation for the analysis of small molecules. At the current time, the complexity of some workflows and the lack of electronic interfaces to some LBA emerging technologies can adversely impact the use of automation in the bioanalysis of protein therapeutics. Automation will not succeed in any organization without a clear strategy with short-and long-term goals, sufficient capital, in-depth training, sufficient time, and dedicated subject-matter experts (SME). In our experience, the most critical factors among the above items are to have the dedicated SMEs and a flexible but systematic plan. These two factors are essential to integrate automation into the culture of an organization and to address the challenges of any process implementation in a regulated environment. Once automation is successfully implemented, it is important to review the lessons learned and evaluate the ROI (e.g., efficiency gains, improvements in data quality, etc.) with the rest of the organization to help build a culture that embraces automation.

There are multiple automation platforms available in the market and different models of automation implemented across organizations. The automation platforms of choice mainly depend on the familiarity of the individuals with the platform and their preferences. In a resource constrained industry environment, many organizations do not have the capital investment available to evaluate multiple automation platforms. Hence, there will not be a discussion on the different platforms in this chapter. We will briefly touch on the different organizational models and the types of practical workflows commonly used for the automation of LBAs.

9.8.1 Centralized Model

This model requires the creation of a dedicated group of individuals who are SME in automation. The key to the group's success is to provide them with the time and tools to implement automation in different workflows, some of which may be converted to "open-access" with the aid of visual tools and training of nonautomation experts.

This is a great model to "jump-start" an automation initiative and demonstrate a ROI to the organization. This model is highly efficient when the sample preparation workflows and the assays are of high complexity.

9.8.2 Open-Access Model

An open-access model is one where automation instrumentation is available to anybody in the organization willing to try it. This type of model can help expand the implementation of automation across an organization if the workflows automated are simple, generic, and require minimal expertize to operate. However, it can lead to disappointment to the novice user trying to implement it in a complex workflow environment, as it may require a considerable time commitment. If the desired results are not as expected, the analyst may go back to manual workflows without a desire to try automation again.

Even in an open-system model, one or two SMEs are often required to oversee the maintenance of the equipment and to aid the less proficient users in troubleshooting and method development.

Some vendors now offer user-friendly interfaces that are ideal for open-access systems as the software displays images of the liquid handler's deck layout, prompts the user to answer questions, and uses the answers to build the final script, requiring minimum to no knowledge of the specific automation platform software.

9.8.3 Challenges and Disadvantages

As mentioned above, time and labor resources are needed for the successful implementation of automation in a bioanalytical laboratory. In addition, it is important to avoid some potential pitfalls. For example, even when multiple liquid handlers are available, unless both systems are "calibrated to match," subtle differences in accuracy between units (even if within specifications) may result in recovery differences that result in only one unit being qualified to support validation and sample analysis.

Another common challenge may be resistance in the organization to automation. Hence, a careful strategy to transform the laboratory's culture is often required. Pairing up "nonbelievers" with automation super users is helpful when implementing this transformation.

During the early stages of implementation, and before instrument redundancy is established, it is recommended to incorporate manual runs in the validation of automated assays, so that sample analysis can proceed even if the instrument is out of service. Table 9.3 shows an example of how accuracy and precision runs may be designed to incorporate both in the validation plan.

Table 9.3 Example of a robust validation design combining several factors including automation[a]

Run ID	Plate lot	Incubation length[b]	Washer used	Reader used
Analyst 1 day 1 manual				
Run 1	1	Middle [60/60/45]	A	A
Run 2	2	Long [75/75/60]	B	A
Analyst 1 day 2 automation				
Run 7	1	Short [45/45/30]	B	B
Run 8	2	Middle [60/60/45]	A	B
Analyst 2 day 1 automation				
Run 3	1	Middle [60/60/45]	B	A
Run 4	2	Long [75/75/60]	B	B
Analyst 2 day 2 manual				
Run 9	2	Short [45/45/30]	A	A
Run 10	1	Long [75/75/60]	A	B
Analyst 3 day 1 manual				
Run 5	1	Short [45/45/30]	B	B
Run 6	2	Middle [60/60/45]	B	B
Analyst 3 day 2 automation				
Run 11	2	Short [45/45/30]	A	A
Run 12	1	Long [75/75/60]	A	A

[a]Designed by Sean Crawford, Bristol-Myers Squibb. Bracketed numbers under the incubation column denote the time in minutes for incubation steps in the method
[b]Bracket time in minutes for an assay with three incubation steps

9.8.4 *Easy Wins*

When introducing the use of automation to a group, it is important to identify simple workflows such as reagent aliquoting and preparation of standards and QCs to achieve some easily noticeable advantages for implementation. These types of applications can later be converted to "open-access" to disseminate their use to the larger workforce. Besides the use of liquid handlers, such as the Tecan, Hamilton or Janus, there are platforms such as the Gyrolab workstation that offer walkaway routine sample analysis, freeing up time for the bioanalytical scientist.

9.8.5 *Types of Automation*

During the drug discovery phase, it is typical to analyze samples at multiple dilutions because PK profiles for predictive use are not available. Automation may alleviate analysts from this labor intensive and possibly error prone step, but script writing can be time consuming and requires the involvement of proficient users. Duo et al. [49] reported the implementation of five major script modules on the

Tecan Freedom EVO liquid handling software to facilitate sample dilution, sample minimum required dilution, standard/QC minimum required dilution, standard/QC/sample addition, and reagent addition. The modular design of automation scripts contained user-friendly interfaces for inputs of assay variables that allow novice users to write the needed script within 10 min.

9.8.5.1 Full-Automation

Automation of all steps from sample tube decapping to data reporting may not be a good return on investment for all applications, laboratories, or even all studies being supported within a group. Below are some workflows or segments of the sample analysis process that greatly improve the laboratory's efficiency when automated, even if their integration to other steps of the method are manual.

Automating a custom LBA from decapping a sample tube to data reporting on most conventional robotic automation platforms can be challenging in the regulated environment. While it is not impossible given sufficient time and resources, automating just the execution of the steps required to perform the assay is commonly achievable. The main disadvantage of automating the entire assay on most platforms is the accessibility of the system to do other activities during a fully automated run [40].

When considering full-automation, several aspects beyond script creation must not be overlooked: (1) samples must be collected in automation friendly tubes, (2) the labels must contain a barcode readable by the liquid handler, and (3) an interface between LIMS and the liquid handler must be created. These points are briefly discussed below and may be considered as stand-alone areas that can be automated to relieve the analyst from the burden of running manual assays.

Use of Automation Friendly Tubes

In our hands, implementation of automation-friendly sample collection tubes that allow the use of capper-decapper instruments can result in up to 66% efficiency gains per run, reduce hand-strain motions and exposure to biological hazards. In addition to those benefits, it ensures that samples and controls are stored in the same tube type. Standardization requires collaboration with the sample operations group and may bring challenges for global studies where central laboratories may not have access to the same distributors as the Sponsor but this is time well spent for studies where sample dilutions are being performed by automation.

Barcoding

At the current time, most of the publications referencing the use of barcode scanners integrated with their liquid handlers were outside of the PK bioanalysis space and were in areas, such as plasma biobanking, blood testing, and gene expression analysis, and only available in the vendor's Journal (Tecan Journal 2007). There are challenges associated with the integration of a barcode reader for the handling of bioanalytical samples (1.4–4 mL size tubes) including: (a) flexibility of the central laboratories and clinical sites to change the label size/type to one that will fit the tube, (b) labels not being thin enough to fit the racks, and (c) careless label application may affect the scanner's ability to read it. There are many automated solutions for tube labeling to avoid the above issues, but they have not been implemented across all central laboratories and sites. In the future, ease of incorporation will come when central laboratories switch to the use of 2D pre-barcoded tubes. It is important to confirm compatibility of these tubes with the scanner as some scanners may require high contrast between the background and the code in order for it to be scanned.

Utilization of Macros to Interface with LIMS
There are no universal requirements across all instrument companies in terms of the format and type of the input (work list) and output (data) files that are compatible with their systems. Watson (Thermo Fisher Scientific, Waltham, MA) is one of the most widely used laboratory integration management systems (LIMS) in the bioanalytical field. However, most commonly used LBA platforms are not compatible or allow the import of work lists from Watson into their software. Additional complications arise when users want to utilize liquid handlers to automate the sample dilution steps. Simple reformatting of the work list may be achievable with an Excel macro as reported by Hao et al. (Tecan application). Alternatively, Dodge [50] took advantage of the Gemini Work List Language (GWL) programming option to control the Tecan liquid handler to write a visual Basic macro that converts the Watson LIMS work list into a GWL file. The GWL commands allows exact control of aspiration and dispensing steps to ensure most accurate and reproducible pipetting schemes and no programming or liquid handler experience is required to use this scheme. In addition, the macro captures the names of the files being converted, to aid in audit trail tracking which is necessary in the regulated bioanalytical environment.

9.8.5.2 Modular Semi-Automation

The modular form of automation is the most common workflow in many regulated laboratories. The different steps of the assay are separated into modules and the scripts are developed so that they can be executed independent of each other. Some examples are: the plate coating, sample dilution, standard, and QC preparation. The writing of the scripts for each of these modules requires the effort of a proficient and experienced user. These scripts contain user-friendly interfaces for simple inputs of assay variables that enables a novice user to execute these in a short period [49].

– *Use of pre-coated plates* A good example of modular automation is the preparation of pre-coated plates. The preparation of large batches of plates with the coating reagent (pre-coated plates) for use in LBAs is not a new practice, but surprisingly not adopted by all laboratories. A liquid handler or a plate washer with a stacker can be used to perform the coating and blocking of plates ahead of the analytical runs. The pre-coated plates can be blocked with a suitable blocking buffer permitting long-term storage of the plates (e.g., stabilcoat). This step eliminates the potential variability associated with daily plate coating and the hand-straining motions from the process, and saves 2–3 h a day from the overall analysis time. Most monoclonal antibodies (mAbs) coated on a solid surface and blocked with appropriate buffers are stable for long periods. Please note that the proper stability experiments are still required for each assay capture reagent.

9.9 Pre-study Validation—General Information

Validation of LBA is a continuous process. Developing a method for the intended purpose, testing the boundaries of the assays and the use of the assay in a routine setting all contribute to the success of the data generated from these assays. The continuum of validation is based on these three main aspects of the assay's life cycle: robust method development, validation of the well-developed method to assess the boundary conditions, and the use of the validated method in a production environment to test the robustness of the assay. As stated previously, careful consideration of a multitude of parameters help to manage this continuum. Once a set of criteria are established for an assay, the validation process will confirm these criteria.

The main validation parameters for any analytical assay are accuracy, precision, selectivity, specificity, and stability. Secondarily, it is essential to understand the dilutional linearity, dilutional accuracy, robustness and ruggedness of the assay [51]. Depending on the complexities of the reagents and the reference material, the evaluation of parallelism may be required. Since LBA are based on binding interactions, it is necessary to understand the affinity and the specificity of the reagents used in the assays (see earlier reference to critical reagents). As discussed in the previous sections, assay reagents, assay format, and the batch size need to be tested during the method development phase and should not rely on testing during the pre-study validation phase. It is best practice that the plates, CDs, or other platforms that are used in the development phase, match those used in the validation experiments, and ultimately mimic the manner in which the samples will be run. It is recommended that partial plates, CDs, and the like not be used in the validation runs. In a typical LBA, the incubation times used for each step of the assay could be critical for unstable protein therapeutics or reagents during automation of the assays. Therefore, prior to starting validation experiments, each method step must be reviewed and carefully considered.

One of the basic questions often asked is, "How many replicates should I use in the assay?" The answer is very simple. The number of replicates to use depends on the robustness of the assay, which is primarily dependent on the quality and effectiveness of the assay reagents. During method development, a thorough evaluation of the number of replicates needed should be performed. Only after this fundamental aspect of the methodology is understood, should the assay progress toward validation. If more than two wells are deemed statistically necessary to measure a sample, then a priori acceptance criteria needs to accommodate the validation experiment design. Irrespective of the number of wells that are used, one average result should be reported as the final concentration value.

Specificity is a parameter tested during the selection of the reagents to assess cross-reactivity to other analytes expected to be present in the sample (e.g., co-dosed drugs); this assessment can be included as part of the validation process. It is a general practice to first test the specificity of reagents in a buffer environment devoid of any complexities that may arise from biological matrix. The specificity of reagents should be tested across the range of the standard curve, as there could be interferences affecting the lower versus upper end of the curve (e.g., a concentration dependent specificity issue; see discussion earlier in this chapter).

Depending on the organizational structure, it may be customary for the person who developed the assay to be involved with validation. However, it is best practice to include additional scientists when running the validation experiments to obtain more robust and reproducible estimates of assay limits. It is best practice that the plates, CDs or other platforms used in the validation experiments mimic the manner that the samples will be run.

9.9.1 Documentation of the Method Procedure

Prior to starting method validation activities, the detailed steps of the method should be documented. The format of the method documentation depends on the company and their practices, but critical parameters, based on method development data, should be noted in the method procedure. Once the assay is validated, these conditions should be finalized and the acceptance criteria established for the method. There should also be a formal process for change control of the method prior to supporting any regulated work, including signatory approvals and a documented audit trail. Though alternative processes may be acceptable in different organizations, the main goal is to "lock" the method, prior to its use in study sample analysis.

9.9.2 Documentation of Validation Experiments

A description of the experiments to be conducted to test each of the validation parameters needs to be written in the form of a validation plan or SOP. This practice may differ depending on the organization but a formal procedure needs to be in place prior to starting the pre-study validation experiments. It is equally important to ensure that all materials and information required are available. In this regard, we offer this sample checklist for use prior to starting the formal validation process:

- A draft analytical method.
- All reagents in sufficient quantities to complete the pre-study and in-study validations including incurred sample reanalysis (ISR) testing. It is preferred to have one lot of critical reagents.
- A suitable lot and quantity of reference material with a Certificate of Analysis.
- Matrix in sufficient quantities to prepare standards, quality controls, stability samples, and for sample dilutions.
- Hardware available as needed—instruments calibrated and maintained and sufficient consumables (plates, CDs, etc.) on hand.
- Analysts with appropriate and documented training.
- A validation plan/protocol or a SOP to guide the validation.

A typical validation run using a 96-well plate format is depicted in Table 9.4.

9.9.3 Accuracy and Precision

Accuracy and precision (A&P) is determined using the QC sample results. The QC samples need to be prepared in the same matrix as the study samples. The QC samples need to be prepared independent of the standard curve and assessed only after a suitable calibration curve is established. In general, three levels of QC samples should be prepared: A low QC sample (LQC) that is $\leq$3x the lowest standard concentration; a medium QC sample (MQC) prepared in the middle of the curve based on a log scale; and a high QC sample (HQC) prepared at a concentration $\sim$75% of the highest standard. Please note that the concentration of anchor points are not used in determining the levels at which the QC samples should be prepared. It is customary to prepare a set of QCs for use during the entire validation. These should be aliquoted into volumes similar to those anticipated in the study samples. Ideally, the quantity of QCs prepared should be sufficient to cover the entire validation experiments as well as stability assessments.

It is extremely important to execute validation experiments in a fashion that reflects the manner in which the method will be used. In that way, the data generated from validation activities will more closely reflect the performance of the method in a production (i.e., sample analysis) setting. The practice of performing validation studies under "best case" conditions may lead to an expectation that the method will perform better in a production environment than what is eventually observed, and is discouraged.

Table 9.4 An example of a typical validation run

	1	2	3	4	5	6	7	8	9	10	11	12
A	Blank	Blank	S8	S8	DILQC1	DILQC1	LLOQ2	LLOQ2	Sel 5B	Sel 5B	Sel 9B	Sel 9B
B	S1	S1	S9	S9	Sel 1B	Sel 1B	LQC2	LQC2	Sel 5	Sel 5	Sel 9	Sel 0
C	S2	S2	S10	S10	Sel 1	SEl 1	MQC2	MQC2	Sel 6B	Sel 6B	LLOQ3	LLOQ3
D	S3	S3	LLOQ1	LLOQ1	Sel 2B	Sel 2B	HQC2	HQC2	Sel 6	Sel 6	LQC3	LQC3
E	S4	S4	LQC1	LQC1	Sel 2	Sel 2	ULOQ2	ULOQ2	Sel 7B	Sel 7B	MQC3	MQC3
F	S5	S5	MQC1	MQC1	Sel 3B	Sel 3B	DILQC2	DILQC2	Sel 7	Sel 7	HQC3	HQC3
G	S6	S6	HQC1	HQC1	Sel 3	Sel 3	Sel 4B	Sel 4B	Sel 8B	Sel 8B	ULOQ3	ULOQ3
H	S7	S7	ULOQ1	ULOQ1	Empty	Empty	Sel 4	Sel 4	Sel 8	Sel 8	DILQC3	DILQC3

9.9.4 Dilutional Linearity and Dilutional Accuracy

Dilutional linearity and dilutional accuracy are two important but different parameters that are often confused within the bioanalytical community. Preparing and analyzing a highly concentrated sample, diluted into the middle of the calibration curve range determines the dilutional accuracy of the method. It is a good practice to include this dilution QC during the A&P experiments to establish A&P parameters for a diluted sample. Once acceptable accuracy and precision is established for the dilution QC, this dilution scheme can be used for diluting high concentration samples into the calibration range eliminating the need to run dilution QCs during in-study runs where samples are diluted into the assay range. While this is not a universally accepted practice, running dilution QCs, in our view, does not add value to the data, though it may add an additional level of comfort that the sample dilutions were performed correctly.

Performing and analyzing a series of dilutions on a highly concentrated sample determines the dilutional linearity of the method or the assay's parallelism characteristics. There may be several intermediate concentrations necessary to evaluate this parameter but as described earlier, it is important to characterize a suitable dynamic range for the assay that functions suitably.

9.9.5 Selectivity

Interference testing is described in detail in a previous section of this chapter and the requirements during validation will be discussed here. The selectivity of an assay should be evaluated for both normal and diseased populations during assay validation. A common practice is to screen a panel of normal individuals' matrix that can be combined into a large pool for use in preparing standards and QC samples. It is necessary to test the pooled matrix against a calibration curve prepared in buffer to assure selectivity prior to making the standards and the quality control samples. The test samples used for selectivity should closely mimic the pre-analytical collection procedures described in the protocol (i.e., venipuncture vs. blood bag collection). Selectivity evaluations should be performed using samples in pairs—one test sample with the blank matrix and the other spiked with the analyte-and preferably assayed in the same run and within the same plate or CD. The individual sample results are then compared to each other to ensure that there are no interferences present. Typically, a concentrated solution of the reference material (50X or 100X) is used to prepare selectivity samples. It is also useful to prepare one additional sample from the matrix pool as a control to ensure the selectivity samples were prepared properly.

9.9.6 Stability

The stability of the protein therapeutic in matrix is typically a physiochemical characteristic of the drug. Relative to the stability of the protein therapeutic prior to administration in a study, information on the proper storage, handling, and administration procedures must be documented and communicated to the toxicology facility or clinic. If additives are necessary for use in sample collection tubes, their effectiveness should be tested and established during method development. This helps to ensure that the stability of the analyte is preserved, and that the additive does not cause interference in the assay.

LBA stability assessments generally measure the ability of the capture and detection reagents to bind the protein therapeutic (i.e., immune reactivity vs. true stability). A LBA method may not be sufficient to definitively confirm that the protein therapeutic is unstable in the matrix. So when employing LBAs, a loss of interaction between the reagents and the protein therapeutic is considered instability. Unlike LC-MS analysis, LBA is not able to provide physiochemical or mass information for the analyte of interest. Therefore, any perturbation that affects the binding properties of the analyte to the reagents is considered a stability event until it can be refuted or confirmed using orthogonal methods. Stability evaluations should be tailored to match the intended use of the assay and its execution. The typical parameters tested are freeze/thaw (FT), processed sample (i.e., MRD on the bench top), and short-and long-term sample storage temperature stability.

Evaluation of FT stability has attracted a lot of attention during recent years [33], particularly the determination of a zero FT cycle. In practice, the bioanalyst is unlikely to receive a sample that is not frozen. Therefore, if it is proven that one freeze thaw cycle does not affect the stability of the therapeutic, it is scientifically justifiable to use a frozen standard curve for subsequent analysis, assuming the protein therapeutic has been proven to be stable under the sample storage conditions employed (both temperature an duration).

The processed sample stability experiments should be designed with the end use in mind. The length of the entire assay process needs to be carefully evaluated before launching into the processed sample stability study. For example, if the method necessitates an overnight incubation to attain maximum sensitivity, it is imperative that there is sufficient knowledge on the stability of the analyte under those conditions of testing. Evaluation of the stability of diluted samples (in matrix) followed by the assay MRD is another example of processed sample stability that should be tested if samples will be diluted (in matrix or buffer) and then stored for a determined amount of time prior to their analysis. Though analyte and sample stability is a drug-specific characteristic, most bioanalytical laboratories replicate stability assessments in each individual laboratory performing the analysis. While this may not be necessary for some stability assessments, such as FT and long-term storage, if the processing conditions are different among laboratories the duplication of work may be necessary. In any situation, it is essential that the stability data be shared when the assays are transferred to other laboratories to provide basic information for how to process samples.

Table 9.5 An example of a robust validation design

	Short	Medium	Long
Capture	Analyst 1	Analyst 2	Analyst 2
Analyte	Analyst 2	Analyst 1	Analyst 3
Detection	Analyst 3	Analyst 3	Analyst 1

9.9.7 How to Set up a Robust Design

The simplest form of a ligand binding assay has three major steps: Capture, analyte incubation, and detection. Hence, if there are three analysts executing the validation, it is necessary that the different incubation times for these steps be confirmed during the A&P evaluation An example is shown in Table 9.5.

9.9.8 Accuracy and Precision Evaluation

There are many publications that define this process very clearly [2, 3]. At the time of performing the validation there are no acceptance criteria established to assess the acceptability of runs. Accuracy and precision estimates for the method are established using the cumulative QC data generated during the pre-study validation A&P runs.

When evaluating A&P runs, the "goodness of fit" of the calibration curves should be the only criterion used to assess the acceptability of the validation run and to determine run acceptance. The target acceptance criteria for evaluating the calibration curves should be set a priori in the validation plan, SOP or be included in the method. A process for rejecting standard calibration points needs to be set prior to the start of method validation. When a calibration curve is performed in duplicate, it is best practice not to reject only one value of a duplicate set, since the average should be used in the assessment of the standard curve. If for any reason a calibration curve point is rejected, the standard curve needs to be reregressed without that calibration point. Once each run performed is deemed acceptable, cumulative data from all of the A&P runs are generated, compiled and assessed using the appropriate statistical calculations to determine the method's precision and accuracy [3].

9.9.9 Setting Acceptance Criteria

There are several ways to set the bioanalytical method acceptance criteria. The authors feel that data obtained from the validation A&P runs should form the basis for setting this criteria. If appropriate robustness is built into the validation process, the results of the A&P runs obtained during method validation should reflect the method's performance in a production environment. The total error (X) of the method is approximated by summing the inter-assay precision and accuracy values derived from the quality control results from all A&P runs with an acceptable standard curve. "X" represents the total evaluable error that is comprised of the imprecision (random bias) and the systematic errors. This single value determines the X in the widely accepted 4–6-X rule, where four of the six quality controls analyzed in an assay run, have to be within X%.

As previously described in this chapter, QC and study samples are run in multiple wells for LBAs. It is common practice to calculate the precision of the replicate results. The intermediate precision of the assay should not be confused with the precision between the concentrations of the replicate wells. The latter reflects the variability of the pipetting technique, while intermediate precision also includes all other sources of assay variation. If there are additional acceptance criteria one would like to set based on the variation encountered while pipetting, this should be done separately during validation. The intermediate or inter-assay precision (%RSD) and accuracy (% deivation) are calculated from the concentration results of the quality control samples from all valid A&P runs. Once the total error criteria for the method is finalized, this value is used to determine the acceptance of other runs used to assess parameters such as selectivity, stability, and dilutional linearity.

9.9.10 Reporting of Data and Finalizing the Assay Acceptance Criteria

Once the data and the acceptance criteria are obtained, the method needs to be finalized and a validation report needs to be prepared. The Workshop Conference Report—Quantitative Bioanalytical Methods Validation and Implementation: Best Practices for Chromatographic and Ligand Binding Assays [52], describes the documentation requirements of a validation report very thoroughly. This manuscript also describes how the final analytical method is documented.

9.9.11 Method Transfer and Cross-Validation Evaluations

As described in the introduction, many LBAs are outsourced to third-party vendors. The reasons for the outsourcing and the timing of such an activity depend on the individual organization. The main criterion that requires attention is that regardless of where the assay is executed, the data should be comparable to data obtained previously. For this reason, it is necessary to perform a successful method transfer and cross-validation of the assay between the laboratories. The selection process for finding a third party vendor is beyond the scope of this chapter. Each organization typically has criteria for evaluating outsourcing partners in conjunction with the finance and global procurement departments.

A successful method transfer requires the originating laboratory and the receiving laboratory to have open minds to the practices and nuances of the other. There are many situations, where the instruments, SOPs and approaches to the work are different between the laboratories. One aspect to consider upfront is the decision as to which set of SOPs will be used in the transfer. During the method transfer process it is a best practice to actively engage personnel from the originating laboratory when performing the methods in the receiving laboratory's facilities. This will allow for the identification of any differences between the laboratories that might affect method performance. It is advisable that a transfer protocol be adopted to provide the details of the method execution steps as well as clear criteria for troubleshooting, the type of validation to perform (partial or full), and the manner in which the data generated will be evaluated. In a typical method transfer, quality control samples as well as incurred samples are evaluated. A statistical evaluation of the results based on equivalence or difference tests, or a confidence interval approach can be used to objectively evaluate the success of the method transfer. At the end of the method transfer, it is a good practice to present the data in a validation report.

There may be instances where a cross-validation of two different methods (LC-MS/MS vs. LBA) to analyze the same analyte may be necessary. During this scenario, there are more variables to consider such as a comparative evaluation of the dynamic range of each assay type and the identification of suitable QC levels that will work for both ranges in the cross-validation experiment. Cross validating LC-MS/MS against LBA is fraught with fundamental analysis variables such as whether the analyte itself can be considered the same. In most cases the LC-MS/MS assay will measure the total (free + complexed) form of the analyte, while the form measured by the LBA depends on the reagents used to capture and detect the analyte. Due to the complexity of the subject, a full understanding of the biology, pharmacology, and bioanalytical formats of both methods need to be evaluated when trying to determine whether such methods are comparable.

9.10 Study Sample Analysis or In-study Validation

9.10.1 Practical Considerations for Setting up Assays for Sample Analysis

It is important to be prepared for the execution of the assays to support sample analysis just as it is necessary to have a plan for validation. Before the study starts, it is generally a good practice to secure a large pool of the qualified matrix to prepare standard calibrators, QCs and for use in dilutions. Likewise, all critical reagents need to be secured in sufficient quantities to support the entire study or program of studies.

9.10.2 *Preparation of Standards and Quality Controls for Sample Analysis*

The bulk preparation of standards is still highly debated within the bioanalytical community. In situations where a frozen standard curve can be used in the assay and sufficient stability data are available, it is recommended that this practice be followed. If there is not pre-study validation experiments supporting stability of frozen calibrators, then the standard curves need to be prepared daily. The QC samples should reflect the study samples as closely as possible and should be prepared and frozen at least for 18 h prior to their use in the assay. QCs should be prepared so that there is sufficient quantity to support the study, accommodate for any failures, and for completing the ISR testing. It is best practice to prepare and test these; and aliquot and store the QCs in the same locations as the study samples as long as long-term stability data are available [53]. If multiple lots of QCs need to be prepared due to instability issues, it is highly recommended that a portion of each QC batch are secured for longitudinal assessment and for qualification of each new QC batch (often referred to as QC bridging experiments).

9.10.3 *Understanding the PK/TK Profile of the Protein Therapeutic*

The availability of estimates for the anticipated concentrations provides additional benefit when executing the assays. These predicted profiles may be derived from previous studies, and allow the bioanalyst to plan sample dilutions that should result in response values within the quantitation range of the calibration curve. This allows bioanalysis to proceed more efficiently during execution of the assay, and aids in the programming of liquid handlers, if available.

9.10.4 *Setting up Assay Runs in the Laboratory Information Management Systems (LIMS)*

There are a variety of Laboratory Information Management Systems (LIMS) used in bioanalysis for setting up, capturing, and reporting runs from method validation and study sample analysis activities. Independent of the LIMS, it is necessary to have the assay plate map prior to starting the sample analysis. An example of an assay plate map is depicted in Fig. 9.11. The practical benefits of having a standard plate map are; ease of programming of liquid handlers, setting up the sample dilutions and efficiency of the operation. It is a best practice to preset the assay acceptance criteria in the LIMS. Typically this is done for the evaluation of the

(a)

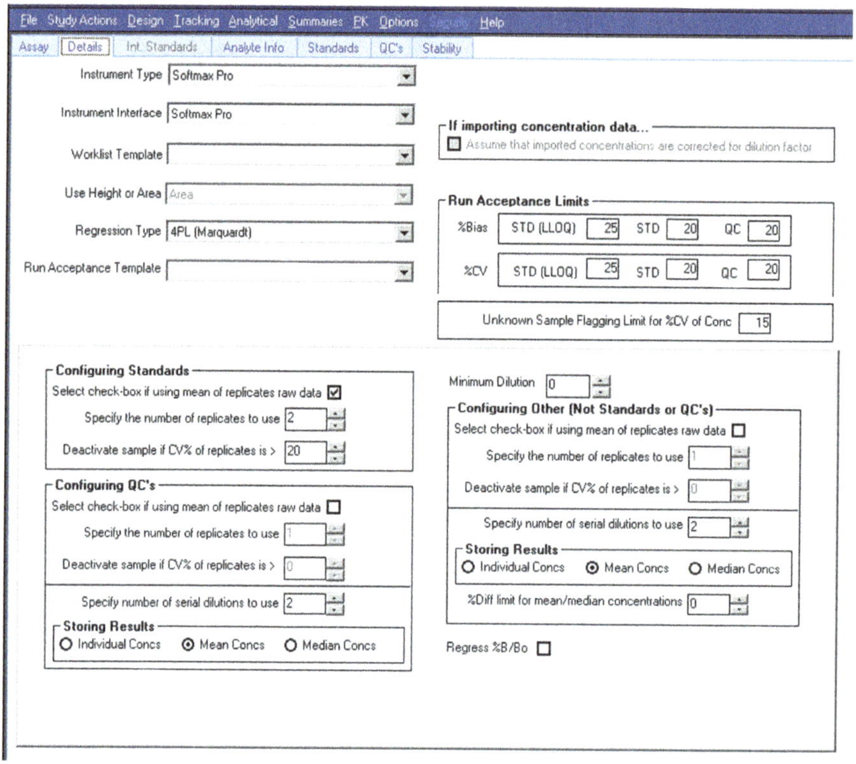

Fig. 9.11 Example of Study Assay (template) set up in Watson LIMS. **a** shows the details for the acceptance criteria for standards and QCs, **b** shows the set up for the standards

standards, quality controls and the samples (i.e., limits of bias, imprecision, etc.) While outliers may be flagged, this may prevent inclusion of their data when creating accuracy and precision tables. There is usually a thorough review of the data by the bioanalyst prior to rejecting a standard and re-regressing the curve, requiring a sample re-assay or completely rejecting a failed run. The LIMS can be used for collating the data at the end of the study, and for trending and monitoring the QC performance during the study.

9.10.5 How to Handle Post-validation Changes

In later stage clinical trials, it is common practice to make changes to the clinical protocols that may necessitate an adaptation to the assay. The common types of changes may include, dosing changes and population changes. When additional

(b)

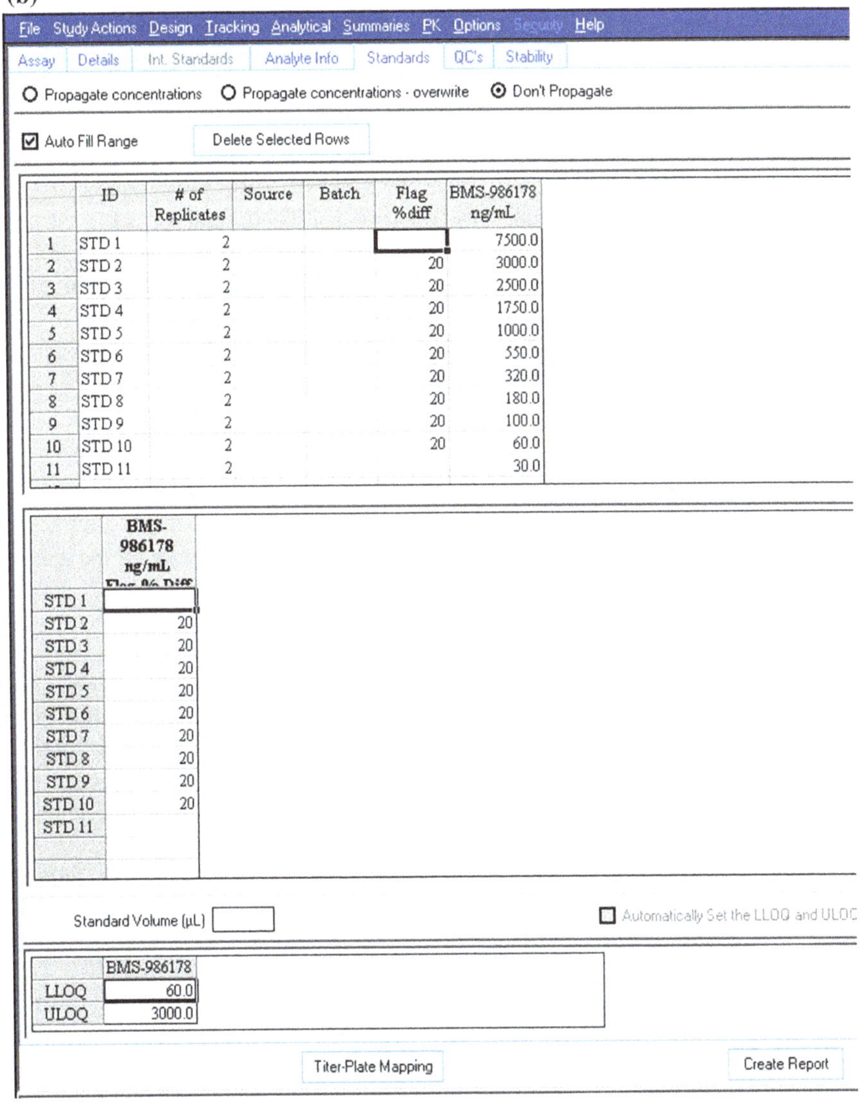

Fig. 9.11 (continued)

disease populations are included in the studies and a selectivity evaluation has not been performed, it will be necessary to retest selectivity. However, if time is limited, permission can be obtained from the clinical groups to use the pre-dose samples from the subjects in the study as part of the selectivity evaluation.

When dosing changes are implemented, additional experiments need to be done to establish dilutional linearity. There are multiple examples where the systemic exposure of the drug (i.e., systemic concentrations in plasma) may exceed the concentrations that can be examined during validation by using the available drug product. This can be caused by multiple dosing and/or the long half-life of the protein therapeutic compared to the available concentration of the drug product. In situations such as this, the bioanalyst will need to use the drug product material available to demonstrate the best possible estimates for prozone effect dilutional linearity, and that dilution linearity from the prepared > HLQ test samples (typically 5% drug product solution 95% matrix) can be accurately achieved. As incurred samples that are beyond the prozone effect concentrations tested in pre-study validation become available, they can be used to further define the limit of potential prozone effects by performing dilution parallelism experiments.

If the LLOQ needs to be changed, i.e., if a more sensitive assay is required then a revalidation of the method will need to be performed.

9.10.6 Incurred Sample Reanalysis (ISR)

ISR testing is another assurance that the method is in control not only for the spiked quality control samples but also in determining the reproducibility of analyte concentration measurements in study samples. There are a number of references on this topic and for LBAs the current industry standard is to run ISR following the recommendations from the Crystal City IV workshop report [54]. The acceptance criteria for LBAs are that at least 67% of the repeated sample results are within 30% of the original value. It is recommended that ISR testing be conducted as soon as possible after completing the study sample analysis. It is also a best practice to run the ISR testing intermittently if the study has a longer duration. This enables troubleshooting of the assay earlier in the course of study sample analysis.

Sample selection criteria, timing, and procedures for evaluating ISR failures need to be established a priori according to the organization's documentation practices. All ISR data should be reported in the respective study reports.

9.10.7 Incurred Sample Stability (ISS)

There is no consensus within the bioanalytical community on the need to perform ISS on a routine basis. As stated earlier in this chapter, LBAs will not provide sufficient information on the actual instability of the biotherapeutic. Since LBAs determine a binding activity (i.e., immunoreactivity of the biotherapeutic to the critical reagents), it will only demonstrate that the biotherapeutic does not yield a response in the assay. While this provides information that the quantification is not possible it will not definitively demonstrate the instability of the molecule.

9.10.8 Incurred Sample Parallelism

The evaluation of parallelism is important for the quantification of endogenous analytes. The parallelism evaluation is done using incurred sample(s) containing the endogenous analytes. There could be situations where an endogenous analyte may not show the same binding properties to the reference material used to prepare standards and QCs in the assay. Therefore, parallelism is not a method issue but an issue with individual samples. While this is not a common phenomenon for a majority of protein therapeutics, when it is evaluated, it is similar to performing a dilutional linearity experiment using incurred samples. The need for doing parallelism experiments depends on the characteristics of the molecule, the endogenous concentration and any known interferences. Strong scientific rationale should justify the evaluation of this parameter during the study sample analysis process.

9.10.9 Sample Chain of Custody

In the Crystal City V meeting, sample chain of custody was discussed as the bioanalytical scientist's responsibility. While not all of the activities from the clinic to the bioanalytical laboratory is under the control of the bioanalyst, every attempt should be made to obtain information related to sample collection, missing time points, storage conditions and shipment information. This practice helps to determine whether any of these variables has the potential to impact the quality of the data.

9.10.10 Documentation of the Study Results

As stated in a previous chapter, all study related data and information is typically documented in a final bioanalytical report. The report and associated documentation needs to be sufficiently detailed to recreate the study if necessary. All sample, ISR, run failures and repeat results need to be provided in an unambiguous format. In addition, calibrator standard and QC data need to be tabulated and appropriate statistical calculations provided for subsequent interpretation by the pharmacokineticist.

9.10.11 Conclusions

Developing, optimizing and validating LBA requires a thorough understanding of the biology of the analyte whether a therapeutic, an antidrug antibody or an endogenous biomarker. Likewise, the stage of the assay's life cycle, the relevant questions that are being asked, and the current regulatory landscape all influence the development, validation, and use of a bioanalytical method. The reagents in an

LBA are key to generating accurate, precise, and robust data. The reagents need to be secured in large quantities once an assay is established and any changes to them need to be thoroughly evaluated and cross-validated during sample analysis.

Ligand binding assay platforms have evolved beyond the plate-based technologies to more complex platforms. While LBAs still remain the workhorse, the use of other technologies will play a significant role in the future.

Acknowledgments The authors would like to thank Sean Crawford for his input in the writing of the DoE section and Lauren Hipelli for editorial help.

References

1. Smolec J, DeSilva B, Smith W, Weiner R, Kelly M, Lee B, et al. Bioanalytical method validation for macromolecules in support of pharmacokinetic studies. Pharm Res. 2005;22(9):1425–31.
2. Findlay JW, Smith WC, Lee JW, Nordblom GD, Das I, DeSilva BS, et al. Validation of immunoassays for bioanalysis: a pharmaceutical industry perspective. J Pharm Biomed Anal. 2000;21(6):1249–73.
3. DeSilva B, Smith W, Weiner R, Kelley M, Smolec J, Lee B, et al. Recommendations for the bioanalytical method validation of ligand-binding assays to support pharmacokinetic assessments of macromolecules. Pharm Res. 2003;20(11):1885–900.
4. Munos B. 2015 New drug approvals hit 66-year high! Forbes. 2016.
5. Yang J, Quarmby V. Free versus total ligand-binding assays: points to consider in biotherapeutic drug development. Bioanalysis. 2011;3(11):1163–5.
6. Williams L, Sank M, Chimalakonda A, Ni Y, Saewert M, DeSilva B, et al. Development and characterization of a free therapeutic ligand binding assay with assistance from kinetics modeling. J Immunol Methods. 2015;419:18–24.
7. Lee JW, Kelley M, King LE, Yang J, Salimi-Moosavi H, Tang MT, et al. Bioanalytical approaches to quantify "total" and "free" therapeutic antibodies and their targets: technical challenges and PK/PD applications over the course of drug development. AAPS J. 2011;13(1):99–110.
8. Bowsher RR, Lynch RA, Brown-Augsburger P, Santa PF, Legan WE, Woodworth JR, et al. Sensitive RIA for the specific determination of insulin lispro. Clin Chem. 1999;45(1):104–10.
9. Mora J, Given Chunyk A, Dysinger M, Purushothama S, Ricks C, Osterlund K, et al. Next generation ligand binding assays-review of emerging technologies' capabilities to enhance throughput and multiplexing. AAPS J. 2014;16(6):1175–84.
10. Fraser S, Cameron M, O'Connor E, Schwickart M, Tanen M, Ware M. Next generation ligand binding assays-review of emerging real-time measurement technologies. AAPS J. 2014;16(5):914–24.
11. Fischer SK, Joyce A, Spengler M, Yang TY, Zhuang Y, Fjording MS, et al. Emerging technologies to increase ligand binding assay sensitivity. AAPS J. 2015;17(1):93–101.
12. Dong H, Mora JR, Brockus C, Chilewski SD, Dodge R, Merrifield C, et al. Development of a generic anti-PEG antibody assay using bioscale's acoustic membrane microparticle technology. AAPS J. 2015;17(6):1511–6.
13. Chilewski SD, Dickerson WM, Mora JR, Saab A, Alderman EM. Evaluation of acoustic membrane microparticle (AMMP) technology for a sensitive ligand binding assay to support pharmacokinetic determinations of a biotherapeutic. AAPS J. 2014;16(6):1366–71.
14. Xu X, Ji QC, Jemal M, Gleason C, Shen JX, Stouffer B, et al. Fit-for-purpose bioanalytical cross-validation for LC-MS/MS assays in clinical studies. Bioanalysis. 2013;5(1):83–90.

15. Thway TM, Ma M, Lee J, Sloey B, Yu S, Wang YM, et al. Experimental and statistical approaches in method cross-validation to support pharmacokinetic decisions. J Pharm Biomed Anal. 2009;49(3):613–8.
16. Gong C, Zeng J, Akinsanya B, Jiang H, Mora J, Chilewski S, et al. Development and validation of an LC-MS/MS assay for the quantitation of a PEGylated anti-CD28 domain antibody in human serum: overcoming interference from antidrug antibodies and soluble target. Bioanalysis. 2014;6(18):2371–83.
17. O'Hara DM, Theobald V, Egan AC, Usansky J, Krishna M, TerWee J, et al. Ligand binding assays in the 21st century laboratory: recommendations for characterization and supply of critical reagents. AAPS J. 2012;14(2):316–28.
18. Haulenbeek J, Piccoli SP. Conjugated critical reagent characterization for ligand-binding assays: using MALDI-TOF-MS as an orthogonal tool to assess assay performance. Bioanalysis. 2014;6(7):983–92.
19. Geist BJ, Egan AC, Yang TY, Dong Y, Shankar G. Characterization of critical reagents in ligand-binding assays: enabling robust bioanalytical methods and lifecycle management. Bioanalysis. 2013;5(2):227–44.
20. King LE, Farley E, Imazato M, Keefe J, Khan M, Ma M, et al. Ligand binding assay critical reagents and their stability: recommendations and best practices from the Global Bioanalysis Consortium harmonization team. AAPS J. 2014;16(3):504–15.
21. Duo J. Comparison of different ligand binding assay platforms for antibody screening against PEGylated therapeutic proteins 2013.
22. Jia Duo AK, David-brown D, Luo L, Haulenbeek J, Liu R, Hamuro L, Zhang Y. Surface plasmon resonance as a tool for reagent screening and characterization to enhance bioanalytical support for biotherapeutic programs 2015.
23. Kozhich A. High throughput screening and pairing of hybridomas at supernatant stage on the gyros. North American Gyros Seminar 2015.
24. Liu RHJ, Krishna M, Duo J, Zhang Y. Anti-Idiotypic Antibody Characterization to support BMS-986090 pharmacokinetics assay 2014.
25. Myler HA, Phillips KR, Dong H, Tabler E, Shaikh M, Coats V, et al. Validation and life-cycle management of a quantitative ligand-binding assay for the measurement of Nulojix((R)), a CTLA-4-Fc fusion protein, in renal and liver transplant patients. Bioanalysis. 2012;4 (10):1215–26.
26. Kaplan IV, Levinson SS. When is a heterophile antibody not a heterophile antibody? When it is an antibody against a specific immunogen. Clin Chem. 1999;45(5):616–8.
27. Lee JW, Devanarayan V, Barrett YC, Weiner R, Allinson J, Fountain S, et al. Fit-for-purpose method development and validation for successful biomarker measurement. Pharm Res. 2006;23(2):312–28.
28. Jones BR, Schultz GA, Eckstein JA, Ackermann BL. Surrogate matrix and surrogate analyte approaches for definitive quantitation of endogenous biomolecules. Bioanalysis. 2012;4(19): 2343–56.
29. Ray CA, Patel V, Shih J, Macaraeg C, Wu Y, Thway T, et al. Application of multi-factorial design of experiments to successfully optimize immunoassays for robust measurements of therapeutic proteins. J Pharm Biomed Anal. 2009;49(2):311–8.
30. Eriksson EJL, Kettaneh-Wold N, Wikstrom C, Wold S. Design of experiments: principles and applications. Sweden: Umetrics AB; 2008.
31. Montgomery DC. Design and analysis of experiments. 3rd ed. New York: Wiley; 1991.
32. Findlay JW, Dillard RF. Appropriate calibration curve fitting in ligand binding assays. AAPS J. 2007;9(2):E260–7.
33. Booth B, Arnold ME, DeSilva B, Amaravadi L, Dudal S, Fluhler E, et al. Workshop report: Crystal City V–quantitative bioanalytical method validation and implementation: the 2013 revised FDA guidance. AAPS J. 2015;17(2):277–88.
34. Sailstad JM, Amaravadi L, Clements-Egan A, Gorovits B, Myler HA, Pillutla RC, et al. A white paper–consensus and recommendations of a global harmonization team on assessing the impact of immunogenicity on pharmacokinetic measurements. AAPS J. 2014;16(3):488–98.

35. Kelley M, Ahene AB, Gorovits B, Kamerud J, King LE, McIntosh T, et al. Theoretical considerations and practical approaches to address the effect of anti-drug antibody (ADA) on quantification of biotherapeutics in circulation. AAPS J. 2013;15(3):646–58.
36. Tate J, Ward G. Interferences in immunoassay. Clin Biochem Rev. 2004;25(2):105–20.
37. DeForge LE, Loyet KM, Delarosa D, Chinn J, Zamanian F, Chuntharapai A, et al. Evaluation of heterophilic antibody blocking agents in reducing false positive interference in immunoassays for IL-17AA, IL-17FF, and IL-17AF. J Immunol Methods. 2010;362(1–2):70–81.
38. Montrose-Rafizadeh C, Yang H, Rodgers BD, Beday A, Pritchette LA, Eng J. High potency antagonists of the pancreatic glucagon-like peptide-1 receptor. J Biol Chem. 1997;272 (34):21201–6.
39. Huang SM, Zhao H, Lee JI, Reynolds K, Zhang L, Temple R, et al. Therapeutic protein-drug interactions and implications for drug development. Clin Pharmacol Ther. 2010;87(4):497–503.
40. Wang J, Patel V, Burns D, Laycock J, Pandya K, Tsoi J, et al. Laboratory automation of high-quality and efficient ligand-binding assays for biotherapeutic drug development. Bioanalysis. 2013;5(13):1635–48.
41. Leung SS, Dreher EA. Automate it: ligand-binding assay productivity in a discovery bioanalytical setting. Bioanalysis. 2013;5(14):1775–82.
42. Allinson JL, Blick KE, Cohen L, Higton D, Li M. Ask the experts: automation: part I. Bioanalysis. 2013;5(16):1953–62.
43. Shen JX. Regulated bioanalytical laboratory automation: where we came from, where we are and where we are going. Bioanalysis. 2011;3(13):1415–8.
44. Li M, Chou J, Jing J, Xu H, Costa A, Caputo R, et al. MARS: bringing the automation of small-molecule bioanalytical sample preparations to a new frontier. Bioanalysis. 2012; 4(11):1311–26.
45. Li M. Bioanalytical laboratory automation development: why should we and how could we collaborate? Bioanalysis. 2015;7(2):153–5.
46. Li M. Laboratory automation: letting scientists focus on science. Bioanalysis. 2015; 7(14):1699–701.
47. Li M. Automation in the bioanalytical laboratory: what is the future? Bioanalysis. 2013; 5(23):2859–61.
48. Ho S. Best practices for discovery bioanalysis: balancing data quality and productivity. Bioanalysis. 2014;6(20):2705–8.
49. Duo J, Dong H, DeSilva B, Zhang YJ. A generic template for automated bioanalytical ligand-binding assays using modular robotic scripts in support of discovery biotherapeutic programs. Bioanalysis. 2013;5(14):1735–50.
50. Dodge R. Using Tecan GWL command programming to achieve automated LIMS TO LIMS sample analysis on the gyros workstation. BMSIARC Conference 2014.
51. Burns DT, Danzer K, Townshend A. A tutorial discussion of the use of the terms "robust" and "rugged" and the associated characteristics of "robustness" and "ruggedness" as used in descriptions of analytical procedures. J Assoc Public Anal. 2009;37:40–60.
52. Viswanathan CT, Bansal S, Booth B, DeStefano AJ, Rose MJ, Sailstad J, et al. Quantitative bioanalytical methods validation and implementation: best practices for chromatographic and ligand binding assays. Pharm Res. 2007;24(10):1962–73.
53. Stevenson L, Kelley M, Gorovits B, Kingsley C, Myler H, Osterlund K, et al. Large molecule specific assay operation: recommendation for best practices and harmonization from the global bioanalysis consortium harmonization team. AAPS J. 2014;16(1):83–8.
54. Fast DM, Kelley M, Viswanathan CT, O'Shaughnessy J, King SP, Chaudhary A, et al. Workshop report and follow-up—AAPS workshop on current topics in GLP bioanalysis: assay reproducibility for incurred samples—implications of crystal recommendations. AAPS J. 2009;11(2):238–41.

Index

M.L. Rocci Jr. and S. Lowes (eds.), *Regulated Bioanalysis: Fundamentals and Practice*, AAPS Advances in the Pharmaceutical Sciences Series 26,
DOI 10.1007/978-3-319-54802-9

www.ingramcontent.com/pod-product-compliance
Lightning Source LLC
Chambersburg PA
CBHW060742150926
51501CB00003B/74
9783319548005